HISTORY OF THE 6th ROYAL BATTALION 13th FRONTIER FORCE RIFLES (SCINDE), 1843–1923

TYPES ENLISTED IN THE REGIMENT.

REGIMENTAL HISTORY
OF THE
6th ROYAL BATTALION
13th FRONTIER FORCE RIFLES
(SCINDE), 1843—1923

The Naval & Military Press Ltd

❖

Published by
The Naval & Military Press Ltd
Unit 10, Ridgewood Industrial Park,
Uckfield, East Sussex,
TN22 5QE England
Tel: +44 (0) 1825 749494
Fax: +44 (0) 1825 765701
www.naval-military-press.com

FOREWORD

THE story of the 6th Royal Battalion, 13th Frontier Force Rifles (Scinde) is well told in this history. Between 1844 and 1923, a period of eighty years, it has served pretty well in every country where Indian troops have been employed. My own connection with the corps dates back to the Mahsud-Waziri Expedition of 1881, forty-four years ago, when, as the 6th Punjab Infantry, it formed part of General Kennedy's column; and I, then a subaltern, little dreamed it would be my good fortune to one day become Colonel Commandant.

In 1908, as the 59th Scinde Rifles (Frontier Force), the Battalion served with me in the Zakha-Khel-Afridi and Mohmand expeditions. A fine spirit animated them, and I was able to see for myself the bond of friendship that united all ranks.

As G.O.C. Northern Army in India, I invariably noted against each corps what I considered its strongest point, and on looking over my diaries I find that against the 59th I always wrote "Very reliable." This they certainly proved themselves to be in France and Flanders, 1914–15, where I was fortunate in having them under my command; and their record there, and later in Mesopotamia and Palestine, is one they may well be proud of.

The Battalion was one of the few units in the Indian Army to be given the title of "Royal" for its services during the Great War, and not one better earned it.

This history is a stirring tale of disciplined valour and sustained loyalty. It will be read by all who have served, or may in the future serve, in the Corps with feelings of pride, and each one as he reads it will have reason to be thankful that he can lay claim to having taken a share, however small, in its honourable career.

James Willcocks
General.
Colonel Commandant.

October **21*st*, 1925.**

CONTENTS

APPENDICES

ILLUSTRATIONS

ILLUSTRATIONS—*continued.*

SKETCH MAPS AND PLANS

The colour plates in this reprint are placed after page 74

REGIMENTAL HISTORY

OF THE

6th ROYAL BATTALION 13th FRONTIER FORCE RIFLES (SCINDE), 1843—1923

CHAPTER I

The Scinde Camel Corps, 1843–1853

Raising of the Regiment.

Major-General Sir Charles Napier, K.C.B., Governor of Scinde, recommended the formation of a Camel Corps, for the purpose of carrying infantry, in a letter dated June 20th, 1843, to Lord Ellenborough, Governor-General in Council. The Camel Corps had been suggested by Lord Ellenborough at the same time that Sir Charles Napier submitted his recommendation for its formation, and was an imitation of the Dromedary Corps which Napoleon had used in Egypt. The special object with which the Corps was raised was to keep in check the wild tribes of Baluchistan and Scinde.

The first Commandant appointed was Lieutenant R. FitzGerald, an officer who was specially chosen on account of his great ability and boldness, his capabilities as a draughtsman, and his ambition to distinguish himself.

The Regiment was raised at Karachi in December, 1843, volunteers being obtained from regiments in the Bombay Presidency. The establishment of the Corps was laid down at 5 subadars, 5 jemadars, 30 havildars, 30 naiks, 10 buglers, and 500 sepoys; 5 first-class and 15 second-class camel jemadars, 30 duffadars, 10 buglers, and 500 camel sowars, or drivers. The camel drivers and camels were obtained from the Marwar and Bikanir districts. It was originally suggested that the camel sowars should be armed with a spear, but they were subsequently armed with short musquetoons, which were made from old muskets cut down. This musquetoon was slung across the back, with the idea that when the sepoys, or riders, were called upon to leave their camels some miles behind in cases of wishing to surprise an enemy, the sowars could form their camels into a circle, heads to the centre, and secure the animals' legs, thus forming a redoubt from which they could offer a stout resistance.

Food for both man and beast, and also bedding, were carried on the saddle, so that no special transport was required. It was the intention that, on nearing an enemy, the sepoys would dismount and advance as infantry, leaving the camels to rest and feed in charge of the sowars. In case of a strong enemy being encountered, both the sepoy and the sowar could get inside squares formed by the camels; and if cavalry attempted to harass the Corps on the march, the infantry-man could load and fire while still riding. It was estimated that the Corps should be able to perform one or even two consecutive marches of sixty miles at the rate of five miles an hour.

From the above it can be seen that a corps was formed which was capable of acting as cavalry and infantry combined, and yet with greater effect than either. Unhampered by baggage of any description, it could undertake long marches and strike with great swiftness. No enemy could get away from them, and in the case of meeting with a superior enemy force, retirement until the arrival of reinforcements could easily be effected. The Corps was therefore most suitable for dealing with the wild inhabitants of Baluchistan and Scinde.

Shortly after the Regiment was raised it was moved to Hyderabad, where it did not remain more than a couple of months before being moved to Larkhana, where lines were built and the Regiment was regularly cantoned. A fort was constructed at Larkhana, which was used as a base for future operations against the mountain chiefs of Baluchistan, who were giving trouble at this time. Shortly after his arrival there, FitzGerald accomplished a march of one hundred miles in forty-eight hours, and succeeded in carrying off a criminal chief from the midst of his tribe, with the result that 115 of the mountain chiefs came

SIR CHARLES NAPIER PURSUING THE ROBBER TRIBES.
The Camel Corps can be seen on the right.
From an old Print

THE SCINDE CAMEL CORPS AS WHEN FIRST RAISED AT KARACHI.

From an old Print.

in to make their " salaams " to the Governor of Scinde immediately afterwards. Sir Charles Napier reported to the Governor-General on March 26th, 1843, that the submission of these chiefs was due to the impression made on them by the Scinde Camel Corps. FitzGerald had accomplished his march at an average rate of five miles an hour, the men carrying 60 rounds of ammunition each, and also bedding. He considered that this could be improved on considerably if orders were given that only the very best camels should be supplied to the Corps.

Operations, 1844-45.

In 1844, operations were undertaken against a hill chief named Baiza Khan. A plan was put forward by Lieutenant FitzGerald, by which he proposed to surprise Baiza Khan in his town of Poolajee. FitzGerald was acquainted with the country, and the plan was agreed to. The Camel Corps, supported by cavalry, was to make a forced march of some sixty miles, and be followed to Poolajee by infantry and guns. However, it was not realized that Poolajee was a walled town, and in the meantime Baiza Khan managed to obtain information of what was being planned. The Camel Corps and the cavalry lost their way in marching across the desert, and arrived at Poolajee at eight o'clock in the morning, much fatigued by their march, before a strongly defended fortress. FitzGerald displayed great bravery in leading his men in an impetuous attack against the city gate, attempting to blow it up with a bag of powder, which was carried by a sergeant, who had effected the same gallant exploit at Ghuznee during the Afghan Campaign. The sergeant, together with ten others, was killed, and twenty-one were wounded. FitzGerald had a most marvellous escape. He was well known to the enemy, and was easily recognizable from the men by his different uniform and his giant stature. The attack on the fortress failed, and the force was compelled to retire to the nearest post, where they were able to obtain water. From here the retreat was continued to Khan Gur (the present Jacobabad), a distance of seventy-seven miles, under a very hot sun. The sepoys showed great endurance during this long march, which was performed without a halt, except to obtain water after leaving Poolajee. Only one sepoy fell out during this trying retreat in very hot weather.

Operations against Baiza Khan were continued again later, when provisions began to run short, and the men were put on very short rations. General Sir Charles Napier therefore despatched the Scinde Camel Corps to Shahpur with orders to scour the country for supplies and return to camp as soon as possible. The following extract from a despatch written by Sir Charles Napier from camp at Truckee on March 9th, 1845, shows the value of the fine march performed by the Camel Corps :—

" On one occasion we were so closely pressed by the scarcity of provisions that I sent off the Camel Corps, under Lieutenant FitzGerald, who reached Shahpur in one march from the Jummuck Pass, making three marches in one, and then returned the same distance with 43,000 pounds of provisions, thus doing in two days and a night what a convoy of hired camels would take six days and six nights to perform, besides requiring a guard ; whereas the Camel Corps required no guard, the sowars being well armed with muskets. There could scarcely be a better specimen of the great power of this Corps even in its infancy."

The operations against Baiza Khan were subsequently brought to a successful issue, and the Camel Corps returned to Larkhana. Shortly after their return to Larkhana, the Regiment made a forced march of some fifty miles in one night in pursuit of a Baluch free-booter named Sobha Khan, whom, with his followers, they succeeded in capturing, taking them unawares when asleep in a village.

The Sikh War.

The Sikhs were now in arms, and the Sutlej campaign had commenced. The Scinde Camel Corps were ordered to move on Ferozepore. They crossed the Indus at Roree Fort, and marched from Bukkur via Bahawalpur to a village Morurote (or Mamdote), a few miles from Ferozepore, having accomplished a march of 450 miles in nine days. They arrived at Ferozepore on December 21st, 1845, the day on which the Battle of Ferozeshah was fought.

During this campaign, two companies of the Regiment served at the siege of Multan. In October, 1846, the Regiment was moved up to Shikarpore, and from there on to Sukkur, and Larkhana, where they remained until 1849.

In October, 1849, the Regiment marched for the Punjab, crossing the Indus at Sukkur and marching to Bahawalpur ; from there crossing the Sutlej, they marched via Multan to a place named Dingu, some miles from Gujrat. On December 22nd they proceeded on to Gujrat, and on January 20th, 1850, arrived at Wazirabad.

Shortly after this the Regiment was ordered to march to Dera Ismail Khan, where lines were to be built, and the Regiment cantoned. The Regiment arrived at Dera Ismail Khan in February, 1850, and lines were built there during that year.

Operations against the Sheranis.

In December, 1851, a detachment of the 5th Punjab Cavalry and a detachment of the Scinde Camel Corps had an engagement with a body of Sheranis who came down from the hills opposite the Draband outpost. In March, 1853, Captain Bruce, commanding, entered the hills in front of Draband with a detachment of 75 men from the Regiment. He was heavily engaged with the Sheranis, and had to retire with the loss of 1 subadar, 4 havildars, 2 naiks and 7 sepoys killed, and 14 men wounded. In April a punitive force, under Brigadier-General Hodgson, operated against the Sheranis, and the Camel Corps took part in this expedition.

In November, 1853, the Camel Corps ceased to exist, and was formed into an infantry regiment, designated the 6th Regiment or Scinde Rifle Corps.

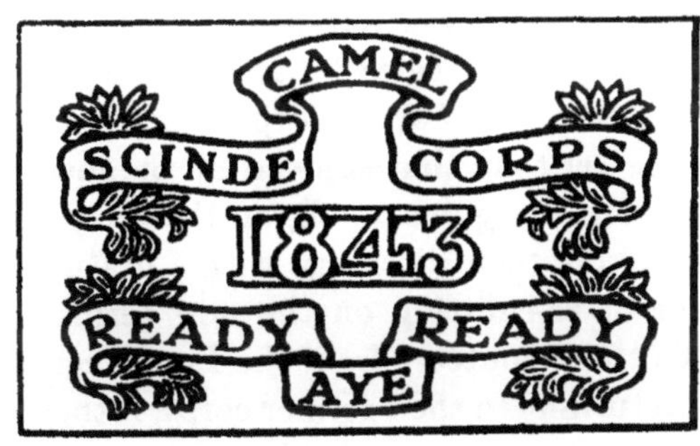

CHAPTER II.

THE 6TH REGIMENT, OR SCINDE RIFLE CORPS, 1853–1856.

General Orders by the Most Noble The Governor-General of India in Council.

"FORT WILLIAM,

September 27th, 1853.

"No. 776 of 1853. Instructions having been issued for reorganizing the Scinde Camel Corps at present stationed at Dera Ismail Khan, as a Light Infantry Regiment armed with rifles, to consist of eight companies, on the establishment and on the allowances of a regiment of Punjab infantry, as laid down in General Orders by the Governor-General, dated February 25th, 1851, The Most Noble The Governor-General in Council is pleased to direct that it shall be incorporated with the Punjab Irregular Force, and designated the 6th Regiment, or Scinde Rifle Corps."

On the first formation of the Scinde Camel Corps in 1843, the soldiers of the Regiment consisted chiefly of volunteers from Bombay Regiments of the Line, and the sowars of the camel establishment were men enlisted originally as Sarwans, but being armed as irregulars and drilled, they were subsequently styled "Camel Sowars."

The pay of the men of the Camel Corps being much superior and the terms of service better than that enjoyed by the men of the Punjab Irregular Force, a choice was given to all, on the reorganization of the Regiment, of either taking their discharge with a gratuity according to length of service, or of re-enlisting on the new terms in the Punjab Irregular Force. To the volunteers from the Line Regiment, a further option was given of returning to their former corps, if they preferred it.

A few only of the volunteers remained, and in accordance with the foregoing General Order they were re-enrolled with many of the young soldiers of the Corps and camel sowars, from November 1st, 1853, from

THE SCINDE CAMEL CORPS COLOURS.

These Colours are hanging in the Garrison Church at Portsmouth, where Sir Charles Napier is buried, and the following is inscribed on a brass tablet beneath the Colours:—

"THE SCINDE CAMEL CORPS."

"The Standards waving above this tablet are consecrated to the Memory of General Sir Charles Napier, G.C.B., the conqueror of Scinde, by whose genius the Scinde Camel Corps was formed. Happier than most conquerors, he secured the affections of the vanquished by a wise and beneficial rule of that noble Province which his valour and military skill had won for his country.

"May his glorious name animate the hearts of British soldiers in the day of battle.

"This tablet is put up by Captain Bruce, Commanding the Camel Corps."

BRITISH AND INDIAN OFFICERS, 1864–65.
Seated (*left to right*) *:* Sub.-Major Rasul Khan, Lieut. J. E. P. Moseley, Lieut. W. C. Chowne, Col. W. D. Hoste. Dr. Constant, Lieut. S. J. Browne, Subadar Mir Sayid.

which date the Regiment was incorporated with the Punjab Irregular Force under the designation of the 6th Regiment or Scinde Rifle Corps.

Scinde Rifle Corps. In G.O.G.G. No. 962, dated November 22nd, 1853, it was ordered that the number be omitted, and the Corps designated simply the Scinde Rifle Corps, and by G.O.G.G. No. 1042, dated August 4th, 1856, the designation of the Regiment was again changed to the 6th Regiment of Punjab Infantry.

1855. In the spring of 1855, the Regiment served with the Field Force under the orders of the Brigadier Commanding, N. B. Chamberlain, C.B., commanding in the Miranzai valley, and returned to Dera Ismail Khan on June 1st. On November 5th the same year, while suffering much from fever, it made a forced march of nearly forty miles to the frontier to repel a threatened attack of the Waziris on Tank. On this occasion the Brigadier was pleased to issue the following order complimenting the Regiment on its alacrity.

Extract from Brigade Orders, No. 13, of November 15th, 1855 :

" Para. 3. The manner in which the Scinde Rifle Corps performed the march of forty miles, with only short halts, is creditable to the Regiment, and when convalescents only lately out of hospital bear up cheerfully against fatigue and keep their place in the ranks, there can be no doubt as to the loyalty and contentment of the Corps."

The Waziris having been dispersed without coming down into the plains, the Regiment returned to cantonments on November 17th.

CHAPTER III.

THE 6TH PUNJAB INFANTRY, 1856–1890.

1856.

ON October 8th, 1856, the Regiment left Dera Ismail Khan to join the field force assembled at Kohat, under the orders of the Brigadier-General Commanding, N. B. Chamberlain, C.B., for services in the Miranzai valley. The Regiment marched with the force on October 22nd, and returned to Kohat on December 22nd, where it was ordered to relieve the 1st Regiment Punjab Infantry.

Outbreak of the Mutiny.

On the first news of the mutiny of the native troops at Delhi and Meerut reaching Kohat, a party of the Regiment, consisting of 1 native officer, 2 non-commissioned officers, and 30 sepoys, was ordered to Khushalgarh to guard the ferry at that place. This guard remained out nearly a month, when it was relieved by a party of levies.

At 10 p.m. on May 22nd, 1857, an order was received from Peshawar directing two companies to be sent to that station without delay. The 3rd and 6th Punjab Infantry each furnished a company consisting of 1 native officer, 9 non-commissioned officers, and 70 sepoys each. These companies marched at 11 p.m., and reached Peshawar (thirty-eight miles) in fifteen hours. The 6th Punjab Infantry company was subsequently employed with Colonel Shute's column in pursuit of the 55th Native Infantry Mutineers, and returned to headquarters on July 8th, 1857.

On June 2nd, 1857, a party of 2 native officers, 6 havildars, 6 naiks, and 80 sepoys, with a party of similar strength from the 3rd Punjab Infantry, the whole under the command of Lieutenant J. Boswell, proceeded by forced marches to Kohat to join Brigadier-General Nicholson's column towards Lahore.

On July 24th a detachment of 4 native officers, 12 havildars, 12 naiks, 4 buglers, and 200 sepoys, marched towards Nowshera for the purpose of relieving the 4th Punjab Infantry, ordered to Hindustan.

On July 12th and 16th the detachment of the Regiment under Lieutenant Boswell, which had joined General Nicholson's column, met the mutineers of the 46th Native Infantry and 9th Cavalry from Sialkot at Trammu Ghat, and charged and captured the Regimental Colours of the 46th Native Infantry. In this action 1 havildar and 2 naiks were killed, and 1 naik and 7 sepoys were wounded.

On August 3rd, 1857, a detachment* under Lieutenant Saunders, forming part of a column operating against the Hindustani fanatics, under Major S. J. Vaughan, attacked and took the village of Noringie in Yusafzai. In this affair 1 sepoy was killed and 1 sepoy severely wounded.

In accordance with instructions received from the Military Secretary to the Chief Commissioner, 2 subadars, 2 jemadars, 12 havildars, and 9 naiks were transferred to the 16th Punjab Infantry at Peshawar. On August 15th, 1857, and on August 21st, 1 subadar, 2 jemadars, 18 havildars, 18 naiks, 2 buglers, and 155 sepoys were transferred to the 9th Punjab Infantry at Kohat.

The Regiment, having now been reduced to its proper complement, was, under the orders of the Chief Commissioner, Punjab, constituted into ten instead of eight companies, each consisting of 1 subadar, 1 jemadar, 6 havildars, 6 naiks, 2 buglers, and 80 sepoys.

The headquarters of the Regiment, and the Right Wing, consisting of 5 subadars, 5 jemadars, and 470 other ranks, under command of Lieutenant T. Quin, marched to Peshawar to join Brigadier-General Cotton's column proceeding into Yusafzai, and on May 5th formed part of the force which attacked and captured the
1858. village of Sitana†. In this affair 1 sepoy was killed and 4 sepoys were wounded. The field force was broken up on May 14th, 1858, and the Wing returned to Kohat on May 20th.

The Left Wing of the Regiment, under command of Lieutenant Saunders, marched to Bannu on September 28th, 1858, to relieve the 3rd Sikh Regiment, ordered to Hindustan. This detachment rejoined at Kohat on December 14th, 1858.

* It would appear from the wording of the old Regimental History that this is the detachment mentioned as proceeding to Nowshera.

† This was the last remaining stronghold of the Hindustani fanatics, who, aided by funds from India, attempted to raise trouble among the border tribes. Their efforts, however, met with small success.

On January 20th, 1859, the Regiment marched to Dera Ismail Khan, in the course of ordinary relief, arriving there on February 2nd. On December 9th the Regiment was ordered to join Brigadier-General Chamberlain's column, proceeding into Kabul Khel country. The column was met at Gundish on December 29th, and the Regiment returned to Dera Ismail Khan on January 15th, 1860.

1860.

On March 12th a party of 3 subadars, 5 jemadars and 248 other ranks, under Lieutenant Saunders, received orders to march direct to Tank, then threatened by Waziris. Within an hour of the order being received the detachment commenced its march, and arrived at Tank, a distance of forty-two miles, in seventeen hours. The detachment rejoined Headquarters on March 18th, 1860.

On March 23rd another detachment, under Lieutenant Saunders, marched to Roree, on the Tank frontier, where it remained until April 8th, when a wing of the Regiment, under Lieutenant Fisher, arrived for the purpose of forming part of a column assembling there preparatory to entering the Mahsud Waziri hills. This column, under Brigadier-General Chamberlain, C.B., marched on April 17th. Lieutenants Stayd and Housford, Her Majesty's 98th Regiment, were attached to the Regiment during these operations. In an action on May 21st the Regiment formed the left column of attack against the Mahsud Waziris. In this attack the Regiment lost one sepoy killed and one wounded.

On March 2nd, 1861, Captain W. D. Hoste arrived and assumed command of the Regiment, to which he had been posted by G.O.G.G. dated February 16th, 1861. In June of the same year the Regiment was reduced to 8 subadars, 8 jemadars, 40 havildars, 40 naiks, 16 buglers and 600 sepoys.

During the whole of 1862 the Regiment was employed in alternately relieving the outposts of Tank, Teetor, Dubbra, Jutta, Goomed, and Manja, with the 5th Punjab Infantry. On December 10th the Regiment marched in course of relief to Bannu, and were located in the fort in that station.

1863, Umbeyla.

On September 20th, 1863, the Regiment was warned to be ready to march on field service at a

Sir Neville Chamberlain's Position on the Ambela Pass.
October 20th–December 16th 1863

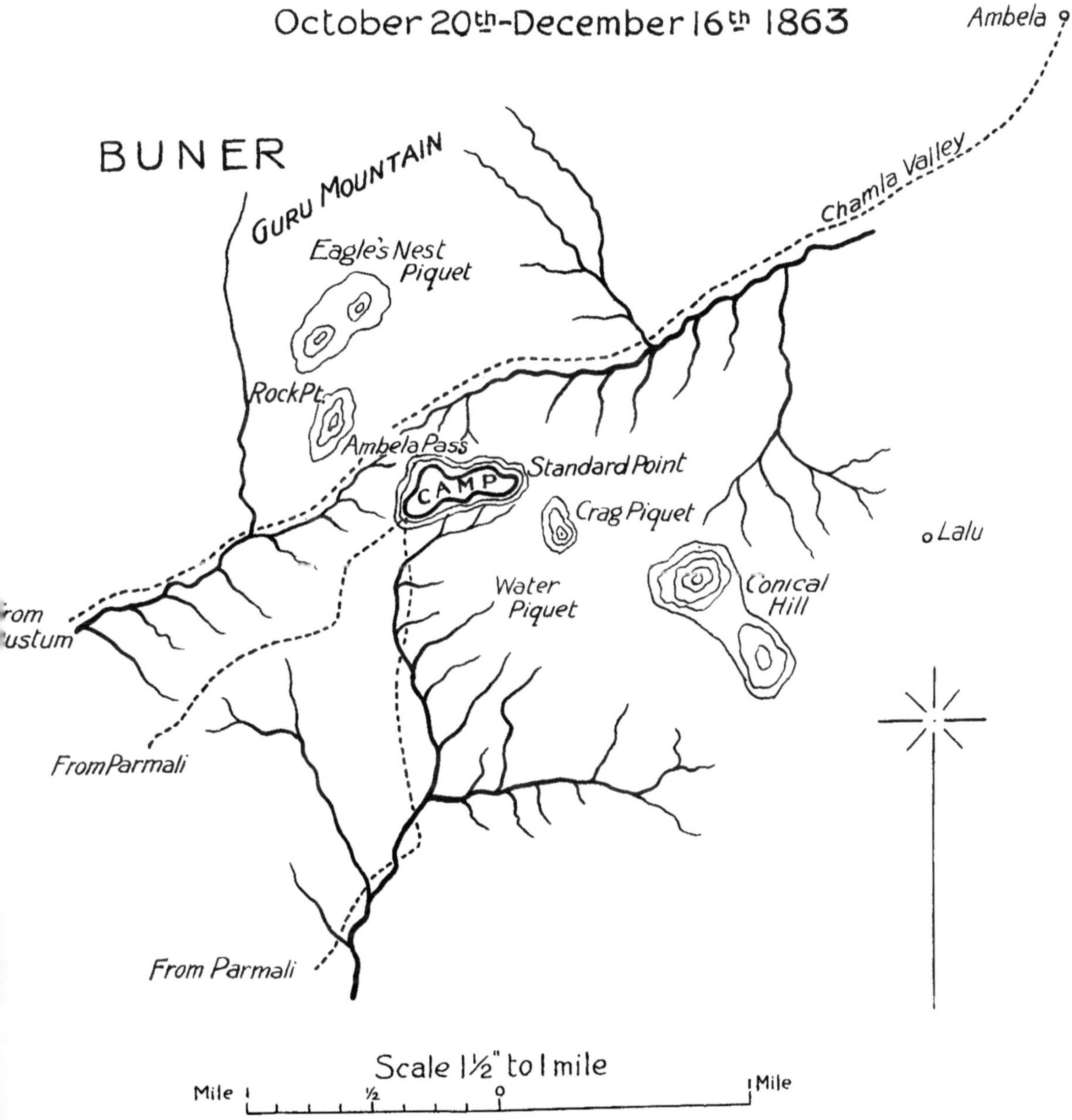

Ambela

BUNER

Guru Mountain

Chamla Valley

Eagle's Nest Piquet

Rock Pt.

Ambela Pass

CAMP

Standard Point

Crag Piquet

Lalu

Water Piquet

Conical Hill

rom
ustum

From Parmali

From Parmali

Scale 1½" to 1 mile

Mile ½ 0 Mile

6/13th F.F.R.

moment's notice, and marched on the evening of the 27th, with orders to proceed to Mardan and join the Yusafzai Expeditionary Force at Nowa Killa, where the Regiment arrived on October 18th. The force broke ground on the evening of the 19th, and reached the memorable heights of Umbeyla on the 20th at 5 p.m.

On the evening of the 22nd a piquet under Subadar Ahmed Khan in the gorge was attacked by the enemy, but the latter were repulsed. Lieutenant Blair, of the Engineers, bore testimony to the gallant conduct of Subadar Ahmed Khan on this occasion.

Early on October 26th the enemy showed in numbers on the Guru Mountain. The Hazara Mountain Train, 200 of the 71st Highland Light Infantry, and the 5th and 6th Punjab Infantry, under Lieut.-Colonel Vaughan, were sent up to the left of Eagle's Nest Piquet to keep them in check. At about 10 a.m. the enemy appeared in force and opened fire on our position. About half an hour later the Regiment charged up at the enemy in skirmishing order and completely routed them. In this affair, Subadars Ahmed Khan and Sirdool Singh, and Jemadars Rasul Khan, Narain Singh and Mehtab Singh were wounded. Havildars Noor Khan and Sunpoorun Singh and 11 sepoys were killed, and about fifty rank and file were wounded.

For conspicuous gallantry on this occasion, the following native officers and men were decorated with the Indian Order of Merit, 3rd Class :—

Subadar (afterwards Subadar-Major) Rasul Khan.
Subadar Mir Sayid.
Jemadar Narain Singh.
Naik Summund Khan.

The Regiment bivouacked that night on the hillside, and were relieved on the following evening by the Ferozepore Regiment.

The Regiment returned to Vaughan's Piquet on the 29th, and held it continually till November 18th, on which date the whole of the picquets on the left of the gorge were abandoned, and the whole force encamped on the right.

The abandonment of the left piquets was effected without the firing of a shot. The Regiment had no sooner reached its ground on the upper camp than it was ordered out with three companies of the

Corps of Guides to form a covering party to the 32nd Native Infantry (Pioneers), who were building the Laloo Piquet, and had a skirmish with the enemy under the Conical Hill, in which the 1st Punjab Infantry took part. One man was killed on this occasion.

On November 20th, when the Crag Piquet was lost for the second time, a party of the Regiment assisted at the recapture, and was most gallantly led by Lieutenant Mackinnon, Officiating Adjutant, who was one of the first in the piquet. Bugler Roor Singh and Sepoy Sarbiland Khan received the Order of Merit for bravery on this occasion.

On December 1st the Regiment moved into the lower camp, having been transferred to the 2nd Brigade, under Brigadier-General A. Wilde, C.B. In the afternoon the Regiment was paraded, when Major-General Garvock, C.B., confirmed the men who had been recommended for the Order of Merit.

On December 4th the Regiment marched to Permoulie, and on the 9th to Sher Durra. From this date they were employed in escorting convoys to the camp. On December 25th the Regiment marched to Nowa Killa, and commenced its return march to Bannu on the 26th, passing through Peshawar on the 30th, and reaching Bannu on January 8th, 1864. During the expedition six officers from other units were attached to the Regiment.

On April 15th, after Brigade parade, Brigadier-General Wilde, C.B., distributed the decorations of the Order of Merit to the men of the Regiment who had earned them by distinguished services at the Umbeyla Pass.

On March 11th, 1866, the Regiment marched to Kohat in course of relief, arriving there on the 17th. On March 11th,
1868. 1868, the Regiment was ordered out at 12 noon, and was led against the Bizotee tribe, as part of the Kohat Garrison, to prevent the enemy from invading our territories.

They first drove the enemy off a conical hill, and were then sent up a precipitous hill in support of the 3rd Regiment. In this action Major Hoste, Jemadar Narain Singh and 23 rank and file were wounded. Sepoy Zarin was awarded the Third-Class Order of Merit.

Officers present in the engagement :—

Captain Quin, who had a most narrow escape, a bullet hitting his helmet in the centre, which must have inevitably killed him on the spot but for the fact that he had just taken off his helmet, and on putting it on his head again happened to leave the chain inside. The bullet, striking against the chain, was turned aside. Captain Browne, Lieutenants Bruce and Moseley, were also present with the Regiment.

Major Hoste and Captain Quin were afterwards thanked by Government for gallantry displayed in the above action. Captain Quin was transferred to the 3rd Punjab Infantry as Commandant on April 20th.

The Regiment marched from Kohat in course of relief on January 2nd, 1869, and reached Dera Ghazi Khan on the 29th.

Brigadier-General Keyes inspected the Regiment on December 19th,
1870, and after the inspection addressed Captain
1870. Browne, who was officiating as Commandant in the
absence of Colonel Hoste on sick leave, and said that it gave him the greatest satisfaction to find the Regiment so steady in their drill and so well disciplined, which showed great pains had evidently been taken, and that he was the more satisfied as he had been at one time so intimately connected with the Regiment, and that he felt sure that Colonel Hoste on his return would be as satisfied as he was, and begged that his remarks might be made known to the native officers and men of the Regiment.

In August, 1871, Enfield rifles were issued to the Regiment.

The Regiment marched from Dera Ghazi Khan on February 3rd,
1872, and reached Dera Ismail Khan on February 14th.
1872. On February 17th the Regiment marched to Zam
to relieve the 5th Punjab Infantry, and returned to Cantonments after completion of outpost duty on March 14th.

On October 28th, at 9 p.m., Lieutenant-Colonel William Dashwood Hoste, the much esteemed and respected Commandant of the Regiment, who had been ill for some time, died very suddenly.

Everyone, from the highest to the lowest, felt the Regiment had suffered an irreparable loss by the death of their Commandant, who had endeared himself to everyone by his genial, frank and kindly manner and the kindly interest and generosity he always displayed in the interests of the men, as also by his gallant bearing in action, during the eleven years and eight months that he had commanded the Regiment.

The following Brigade Order was issued by Brigadier-General C. P. Keyes, C.B., Commanding the Punjab Frontier Force, on November 2nd, 1872 :—

"It is with feelings of very deep regret that the Brigadier-General announces to the Force the death on the evening of the 28th ultimo of Lieutenant-Colonel W. D. Hoste, Commandant, 6th Punjab Infantry.

"Colonel Hoste joined the force in 1861, was selected for the command of the 6th Punjab Infantry by Sir Neville Chamberlain on account of the brilliant reputation he had made for himself on many fields in the Punjab, during the earlier part of the Mutiny with the 5th Punjab Infantry, and latterly with Brigadier Horsford's Column in Oudh.

"Since he had held command of the Corps he had twice led it into action.

"Of the many regiments which added lustre to the British arms on the heights of the Umbeyla Pass, few showed themselves more worthy of the confidence of the Government than did the 6th Punjab Infantry under his gallant leadership; and it was at the head of his men, trying to retrieve a repulse in a skirmish in the impracticable hills near Kohat, that he received that wound to which may in part be attributed his early and much-to-be-lamented death.

"By the Regiment he has so long and as honourably commanded, by English and native officers, by the men of every grade and every class, his loss will be severely felt.

"Few commanders possessed in a higher degree the happy tact of winning the confidence and esteem of all with whom he was connected, and in condolence with the 6th Punjab Infantry on the untimely removal from its midst of one who was one of its best and truest friends, Brigadier-General Keyes feels sure that Colonel Hoste's name will always be remembered by all with sincere regret and loyal affection."

Major B. R. Chambers, second-in-command 3rd Sikh Infantry, was appointed Commandant on January 3rd, 1873,
1873. and assumed command on February 27th of that year.

On March 25th and 26th the Regiment was inspected by General Keyes, and in his remarks on the inspection he was pleased to express

BRITISH OFFICERS AT BANNU, 1864–65.

***Standing*:** Dr. Constant, Colonel L. B. Jones, Third P.C., Lieut. Heame, Assistant Commissioner, Lieut. T. F. Bruce.
***Centre*:** Dr. Rouse, Third P.C., Colonel Johnstone, D.C., Lieut. Browne, Major Salt, R.A., Lieut. T. C. Plowden, Third P.C.
***In front*:** Lieut. Chowne, Lieut. Lockwood, Third P.C., Colonel W. D. Hoste, Lieut. James, R.A.

BRITISH AND INDIAN OFFICERS, EDWARDESABAD, 1875.

Names of British officers, from left to right, are : Surgeon Bookey, Captain T. F. Bruce, Lieut. Yaldwin (*standing*), Major Chambers, Captain Sandilands. S.M. Rasul Khan is seated on the right.

his opinion that much credit was due to Captain S. J. Browne, second-in-command, for the manner in which he had conducted the duties of the Regiment during the time he had held officiating command.

The Regiment was stationed at Dera Ismail Khan during the whole year, and furnished one division of the outposts, the relief being quarterly and the Regiment taking first and second division alternately. The Regiment also furnished a detachment at Sheikh Bodeen from August 15th to the end of October.

The first year of which there are any tabulated records of recruiting is 1873. A total of 106 recruits were enlisted during the year, and of these 68 were enlisted at Regimental headquarters, consisting of 15 Punjabi Mussalmans, 11 Sikhs, 5 Dogras, and 35 Pathans. The remainder—19 Sikh recruits from Amritsar, and 21 Hindustani recruits from the Fyzabad district, were brought in by recruiting parties. From the old Long Roll which was started in 1849, it appears that all the above classes were enlisted from the beginning, and no change was effected until 1890, when an additional company of Dogras was substituted for the company of Hindustanis. The Pathan companies at this time contained many Trans-Frontier men, and even Afghans from Kabul and Jellalabad.

On January 9th, 1874, the Regiment marched from Dera Ismail Khan to Paniala, about thirty miles from the station, near the Sheikh Bodeen Mountain and at the mouth of a large valley, to take part in a camp of exercise, which was composed of the 2nd and 5th Punjab Cavalry, Nos. 2 and 3 Punjab Light Field Batteries, the 1st and 3rd Sikh Infantry, and the 1st and 6th Punjab Infantry, the whole being under command of Brigadier-General C. P. Keyes, C.B., commanding the Punjab Frontier Force.

The camp was broken up on January 30th, and the Regiment returned to Dera Ismail Khan, arriving there on February 2nd. The same outpost duties devolved upon the Regiment in 1874 as in the preceding year.

On January 29th, 1875, the Regiment marched from Dera Ismail Khan in course of relief, and reached Edwardesabad, in the Bannu Valley, on February 3rd.

The outpost duty that was furnished throughout the year by the infantry portion of the garrison at Edwardesabad consisted of 1 native officer and 22 rifles at Latumbar, a post on the high road leading to Kohat, about eighteen miles from that station. One native officer and 34 rifles at Jani Khel Post, situated about fifteen miles from the station in a south-westerly direction, and a post five miles from cantonments to the north-west, where the Kurram River comes out from the hills, which was garrisoned by 1 native officer and 22 rifles. The two former posts were relieved monthly, and the latter every week. The duty was divided between the two infantry regiments at Edwardesabad by one taking the Jani Khel Post, and the other taking the Latumbar and Kurram Posts every alternate month. In addition to this the garrison furnished a detachment of 2 native officers and 68 rifles at Sheikh Bodeen Sanatorium from May 1st to August 15th in the summer, each regiment supplying the detachment for one-half of the above period.

In the year 1876, the station of Edwardesabad was visited by a cholera epidemic during the months of October and November, and the Regiment lost from this fatal disease 1 havildar, 1 naik, 2 buglers, and 7 sepoys, besides 2 camp followers, 4 women, and 2 children. On October 25th the Regiment was marched out to Adumee, some eight miles from cantonments, and remained in camp until November 18th, when they returned to quarters, the disease having disappeared.

Snider rifles were issued to the Regiment on March 25th, 1876. On December 1st the Regiment was inspected by General Keyes.

1877, Jowaki Campaign.

The Regiment was in camp from February 28th to March 16th in the Bannu District for exercise and field firing.

On the night of August 25th, in accordance with orders received by telegram, the Regimental headquarters, 300 bayonets strong, marched for Kohat, and reached Lachi in three marches, distance fifty-two miles, at 2 p.m. on the 28th.

On the afternoon of the 29th the Regiment marched from Lachi reaching Gundiala before daylight. From this place they formed part of the column entering the Jowaki hills, and took their share of the operations carried out that day against the Jowaki tribe. They then retraced their steps to Gundiala, and thence marched to Gumbat, which was reached at 4.30 p.m. on the 30th, the Regiment having been twenty-six hours and a half under arms, and having covered forty miles of ground. Three out of six of the British officers with the Regiment lost their horses, the animals having died from the extreme heat and excessive fatigue.

From September to November 8th, the Regiment was employed on blockade duty on the road between Kohat and Khushalghar, and had a most arduous and harassing time of it, a constant system of patrolling the road having to be kept up, besides detaching small parties during the night into all the adjacent villages.

On November 7th, Lieutenant Vaughan joined headquarters with a detachment of 100 rifles from Edwardesabad.

On November 9th the Regiment marched into the Jowaki hills with the force under the command of Brigadier-General C. P. Keyes, C.B.

No very active operations were carried out for the remainder of the month. In the retirement from the Paiah Valley on November 15th one man was slightly wounded. The Regiment was encamped along a ridge in the Shindi Valley.

On December 1st the Regiment formed part of the force advancing under General Keyes, which attacked and took Jamu, returning to camp at Shindi the same evening.

On December 7th the Regiment formed part of the force which attacked and took Ghariba, leaving Shindi in the early morning, and returning in the evening.

On December 31st the Regiment marched with the force advancing under General Keyes towards Pastaoni, and on reaching Ghariba, the Regiment was told off to form part of a column left at that village to bivouac for the night.

At 2 p.m. on January 1st the force which had advanced on the previous day under General Keyes, commenced retiring from the higher ranges and orders were issued for the 6th Punjab Infantry to cover the final retirement of the whole force from the Ghariba Valley. Dispositions were made accordingly.

At about 2.30 p.m., heavy firing was heard on the left in the Durgaie Pass, and the enemy were perceived to be following up the last regiments returning from the higher range. The force effected its retirement from Ghariba between three and four o'clock p.m., while the enemy were held in check by the 6th Punjab Infantry. A brisk fire was kept up on both sides, but no casualties occurred, nor was the Regiment followed up after the Paiah Valley had been entered.

The Regiment returned to its standing camp at Shindi at seven o'clock p.m.

On January 15th the Regiment formed part of the force which moved out under General Keyes, ascended the Durgaie Ridge, and co-operated with troops which had moved out from Peshawar. Operations were carried out against the enemy on the 15th, 16th, 17th, and 18th, and on the afternoon of the 18th the Regiment returned through the Narkula defile and Jamu, to Shindi.

No further military operations were carried out, and the Regiment remained at Shindi until the Jowaki country was evacuated on March 7th, 1878, upon which date headquarters and the Right Wing commenced its march for Abbottabad, which station was reached on March 22nd.

The left half Battalion marched from Edwardesabad on March 12th, and joined headquarters at Abbottabad on the 30th of the month. The Regiment relieved the 5th Punjab Infantry at Abbottabad and occupied the right infantry lines, the other infantry regiment in the station being the 5th Gurkhas.

At that time there were two outposts in the Hazara District which were garrisoned by infantry detachments. One was at Haripore, where there is an old Sikh fort, in which the *Tahsil* is built, the remainder of the fort being allotted for the accommodation of the detachment, which consisted of three non-commissioned officers and 12 sepoys. The other outpost was at Oghi, in the Agrore Valley, lying at the foot of the Black Mountains, about forty miles (three marches) from Abbottabad. This outpost was a small stone built fort, erected after the Black Mountain Expedition in 1868. The police station was also located in this fort.

In the beginning of this year the strength of this detachment was

1 native officer and 30 rifles, but as the border was somewhat disturbed the strength was increased to 1 native officer and 75 rifles. The Regiment furnished one detachment every month, taking each outpost duty in alternate months.

Early in October the 5th Gurkhas marched on service, after which all military duties in cantonments and district fell upon the Regiment.

On May 30th new accoutrements of the Yoke pattern were received to replace the old kind.

The Regiment was inspected by Major-General Roberts, V.C., on May 17th and 18th.

On November 26th orders were received to recruit the Regiment up to a strength of 800 sepoys. Numerous recruiting parties were at once sent from Headquarters, and efforts were made to complete the increased strength of the Regiment as soon as possible. By February 1st, 1879, the Regiment had recruited up to the new strength.

On July 2nd, 1879, the 2nd Punjab Infantry arrived in Abbottabad, and shared the duties of the district with the Regiment.

Under orders received from the Punjab Government, the Headquarters of the Regiment, with 500 rifles, marched from Abbottabad for the north-west of the Hazara border and encamped on the Chuttar plain, about forty-six miles from Abbottabad, having marched from Cantonments on July 31st.

The Chuttar Plain is a high-lying valley on the border of the Nundihar territory. It is over 5,000 feet above sea level, and is very picturesquely situated. There is a plentiful supply of fresh water, and the climate is very salubrious.

During the rainy season in August and the greater part of September fever was very prevalent amongst the British officers, as well as amongst the men, but it was not of a severe type.

The object with which the Regiment was moved out to the border was as a precautionary measure to hold the Ullahoi tribe in check. The Regiment made a number of reconnaissances in the Nundihar, a valley towards Ullahoi, and in October a reconnaissance with 200 rifles was made along the ridge of the Bhiste and Pulaiga range of mountains to Mulki.

The party was out for five days, and experienced very severe weather, as the range over which the road lay was between ten and eleven thousand feet high. Snow fell daily, and there was a severe frost at night.

In November the cold was very severe during the nights at Chuttar, the thermometer sometimes registering 18 degrees of frost. It was a particularly dry, bracing cold. The men were warmly clad, a good supply of firewood was to be had, and the health of the Regiment was good.

The Regiment returned to cantonments from Chuttar on December 3rd.

The training of recruits who had been enlisted to bring the Regiment up to the new strength was pushed on as rapidly as possible, and they were passed into the ranks in a much shorter time than usual.

On July 12th, 1879, orders were received to stop recruiting, and the Regiment was gradually reduced to the old strength. However, on September 28th, in consequence of hostilities being undertaken in Afghanistan, the increased strength of 800 sepoys was ordered to be maintained, and recruiting was again opened.

It was found that in the end of 1879 men did not come forward for enlistment nearly as readily as they did in the end of 1878. Probably the severe drain there had been on most districts of the Punjab in the previous year told upon the following season.

1880.

Early in February orders were received for the Regiment to prepare for service in Afghanistan, and on March 4th they were inspected by General Godby prior to leaving quarters. The Regiment marched the next morning, 770 strong, and reached Peshawar on the 14th of the month.

After a few days' halt there it moved on to Jamrud on the 19th, but on arrival in camp orders were received countermanding the Regiment going on service, and directing it to proceed to Mardan to garrison that cantonment, where it arrived on the 27th of the month.

In addition to the very grievous disappointment it was to all ranks not being permitted to proceed on active service, this move entailed a very heavy pecuniary loss upon the men, for prior to leaving

Abbottabad they were obliged to sell all their property for what it would fetch, and the married men had to incur the additional expense of sending their families to their homes.

A few days after arrival at Mardan two companies were ordered to march for Abbottabad, and a third company was detached for a couple of months at Attock and employed on escort duty on the road to Peshawar.

On August 19th Headquarters marched for Abbottabad, leaving four companies at Mardan, and reached there on the 27th. One company followed from Mardan a few days later, and the remaining three joined Headquarters on September 22nd. The Regiment remained at Abbottabad until December 22nd, when it marched in course of relief to Kohat, where it arrived on January 5th, 1881.

In the month of March, 1880, orders were published authorizing the enlistment of recruits with bounty. Under this system recruits were enlisted for a term of three years, getting a bounty of Rs50, a portion of which was paid them on enlistment, and the remainder on the completion of their service. This order remained in force until September 1st, and 78 recruits were enlisted under its provisions.

On October 12th recruiting was stopped, and the Regiment was ordered to reduce gradually to the old strength of 640 sepoys.

1881, Operations against Mahsuds.

In March, 1881, the Regiment was warned for service, and on the 29th of that month left Kohat to join the force being formed at Tank, on the Dera Ismail Khan Frontier, under Brigadier-General T. G. Kennedy, C.B., for the purpose of operating against the Mahsud Waziris.

The Regiment reached Tank on April 10th, their strength in the field being 9 British officers, 13 native officers, and 570 other ranks. The 6th Punjab Infantry formed part of the 2nd Infantry Brigade, under the command of Lieut.-Colonel B. R. Chambers, the Regiment being commanded during the expedition by Lieut.-Colonel S. J. Browne.

The force broke ground at Tank on April 18th, and crossed the border on the 22nd, encamping at Jandola. From there the road through the Shahur Pass was taken, and a considerable extent of the Mahsud country was marched over which had never previously been visited by our troops. Camp was pitched on May 5th at Kani Guram.

The force then advanced on to Makin, and from that place returned to Tank via Jandola, arriving on May 15th. Only a slight resistance was offered by the Mahsuds while their country was being occupied. The only casualties suffered by the Regiment during the expedition were two men wounded in some skirmishing at Tangi Ragza on April 28th.

There was a good deal of rough, hard work in flanking and rear-guard duties, and the men were often under arms for a long stretch. On one occasion, on the march from Narai Ragza to Koondiwan, the troops struck tents between four and five on the morning of April 30th, and the 2nd Brigade, which protected the baggage and formed the rear-guard, did not reach camp until two o'clock on the following morning. Many of the marches were very trying to the men, owing to the beds of streams having to be used as roads, which necessitated marching over rough boulders and constant wading through water.

The climate in the Mahsud hills was very pleasant, and the health of the Regiment was good throughout the expedition. One man died the day before the Regiment reached Tank.

Free rations were granted to the troops during the time they were across the border.

On arrival at Tank the force was broken up, and the Regiment reached Kohat on May 31st, the return march having proved a hot and trying one.

Throughout the years 1882 and 1883 the Regiment remained in garrison at Kohat. In April, 1882, the strength was increased by one British officer and 80 sepoys, making the total of all native ranks up to 832. Towards the end of the year there was a considerable amount of sickness, and at its close the Regiment passed through a severe epidemic attack of pneumonia, which continued up to March of 1883. As the result of the epidemic of pneumonia, the Regiment lost thirty non-commissioned officers and men.

On April 1st, 1883, Colonel B. R. Chambers, Commandant, vacated command of the Regiment, and was succeeded by Lieut.-Colonel S. J. Browne. Subadar Narain Singh was appointed Subadar-Major vice

BRITISH AND INDIAN OFFICERS AT EDWARDESABAD, 1886.

Standing : Jem. Akbar Khan, Lieut. Taylor, Sub. Sarbiland, Sub. Bhola Singh, Major Sandilands, Sub. Narain Singh, Lieut. Minchin, Sub. Mowaz Khan, Jem. Subha Singh.

Sitting : Jem. Jelabudin, Sub. Jaipal Singh, Sub. Ghulam Din, Surgeon Major Bookey, Lieut.-Colonel T. F. Bruce, Sub. Bry Basi, Sub. Sandat Khan, Sub. Ghulami, Jem. Khazana.

In front : Lieut. Erskine, Lieut. MacMullen.

VI PUNJAB INFANTRY ON PARADE AT EDWARDESABAD, 1886.

Jani Khan, invalided with effect from May 1st. Subadar Devi Singh, a most exemplary native officer, died on September 1st. He was a great loss to the Regiment.

The Regiment left Kohat for Bannu on February 8th, 1884, where they were ordered to relieve the 2nd Punjab Infantry.

On April 1st, 1884, Class Companies were formed under the authority of Government of India, Military Department, letter No. 28B, dated January 10th, 1884. The composition of the Regiment remained unchanged, the companies being classed as under :—

"A" Company	...	Manja Sikhs.
*"B" Company	...	Mixed Pathans, Yusafzais, Afridis, etc.
"C" Company	...	Dogras.
*"D" Company	...	Bangash Pathans.
"E" Company	...	Hindustanis.
*"F" Company	...	Khattaks.
"G" Company	...	Malwa Sikhs.
"H" Company	...	Punjabi Mohammedans.

In February, 1885, the Regiment was inspected by General Kennedy, and on March 31st a warning was received to be prepared for active service with the 1st Army Corps. The Regiment was afterwards transferred to the 2nd Army Corps. Colonel S. J. Browne was recalled for active service, and rejoined on April 30th, when Colonel Stewart (Q.V.O. Corps of Guides, Officiating Commandant) returned to his regiment. The Regiment was released from "warned for service" in a letter dated October 1st, 1885.

The Regiment remained at Bannu during the whole of 1886. The autumn of that year was an exceptionally healthy season.

The Regiment was inspected by Brigadier-General Sir Charles MacGregor on February 27th.

On January 10th, 1887, the Regiment marched from Bannu, and arrived at Dera Ismail Khan on the 16th of the month, and occupied the left Infantry lines, relieving the 1st Sikh Infantry.

1887.

* The composition of the Regiment at the end of 1885 shows Pathans as being "Trans-Indus, within the border"; "B" Company, Yusafzai; "D" Company, Bangash; and "F" Company, Khattaks. From this it would appear that "B" Company was subsequently made pure Yusafzai, instead of as noted above. The Hindustanis were mostly from Unao.

On February 14th the sad news of the murder of Captain E. B. J. Vaughan, Wing Officer, on special duty in Burmah, at Khany Khuyat, while out shooting alone, was received by telegram, to the deep regret of all ranks, who sorrowed for the loss of a true and gallant officer and comrade.

On March 1st the Regiment was inspected by Brigadier-General J. W. McQueen, C.B., A.D.C., commanding the Punjab Frontier Force.

On May 29th Lieut.-Colonel T. F. Bruce retired from the Service, after an association with the Regiment of nearly twenty-four years. Lieutenant Taylor reported his departure to join the Burmah Commission on December 21st, and was succeeded as Adjutant by Lieutenant Hutchison.

In April, 1887, the strength of the Regiment was raised from 720 to 800 sepoys, and recruiting parties were sent out. Recruits of a good stamp were obtained, with the exception of the Hindustanis.

Throughout the year 1888 the Regiment remained at Dera Ismail Khan. On January 11th Subadar-Major Narain Singh died. His death was felt as a great loss to the Regiment, in which he had served with distinction for over thirty-four years.

Surgeon-Major J. T. B. Bookey returned from field service in Burmah on January 16th, and resumed medical charge of the Regiment. He subsequently was appointed to the command of a Field Hospital in the Hazara Field Force on September 13th, rejoining once more on December 1st.

On June 18th Captain H. B. Urmston was killed in action on the Black Mountain. Captain Urmston had left Aghi Fort early in the morning with a reconnaissance party some eighty strong, under Major Battye, 5th Gurkhas. While still within British territory the party was fired on ; as the advance continued the fire became heavier and the enemy more numerous, and a retirement was ordered. A havildar in the rearguard having been wounded, the two British officers went back to his assistance, and while engaged in putting the wounded man on a stretcher, were charged by the enemy. A hand-to-hand conflict ensued, and both officers were killed. The main body, unaware of what had happened, continued the retirement. An Indian officer, who had been with the rearguard, although himself wounded, managed to rejoin the main body, and led them back and recovered the bodies of the officers.

As a result of this incident, a punitive expedition was despatched against the Black Mountain tribes.

During 1889 the Regiment remained in Dera Ismail Khan, and was inspected by Brigadier-General Sir John McQueen, K.C.B., on February 19th.

The usual malarial fever commenced in September, and though not of a bad type, weakened the men considerably, and made the effects of pneumonia, which started in November, serious. Twenty-five deaths occurred during the year. Among those who died of pneumonia was Subadar Sadhu Singh, whose death occurred on December 16th.

On February 18th, 1890, the Regiment marched from Dera Ismail Khan, and relieved the 2nd Sikh Infantry at
1890. Edwardesabad on the 25th of the same month.

Colonel S. J. Browne vacated command of the Regiment after seven years' tenure on April 1st, 1890, and handed over to Lieut.-Colonel A. N. Sandilands.

Orders were received under Government of India Military Department No. 1420/B, dated May 13th, 1890, to the Adjutant-General in India, sanctioning the caste composition of the Regiment being changed by the substitution of a company of Dogras for the company of Hindustani Hindus, the change to be made gradually.

Captain J. M. Hutchinson, Wing officer, rejoined from duty in Russia (having passed the Russian interpreter's test in London on October 21st, 1890) on December 23rd.

The Regiment formed the only infantry part of the garrison at Edwardesabad from November 3rd until the end of the year.

The health of the Regiment seems to have suffered during its three years' stay at Dera Ismail Khan, and there was an unusually large number of admission of fever cases to hospital at Edwardesabad during the year. The deaths from pneumonia from January 1st to the date of leaving Dera Ismail Khan were five, and from that date to the end of the year at Edwardesabad eleven.

Brigadier-General Sir W. S. A. Lockhart, K.C.B., C.S.I., inspected the Regiment on March 5th, 1891. On March 18th the 4th Punjab Infantry marched into Edwardesabad and relieved the Regiment of some of the heavy duties it had hitherto performed as the sole infantry unit in the garrison since November 3rd, 1890.

CHAPTER IV.

THE 6TH PUNJAB INFANTRY, 1891–1903.

1891, Operations in the Miranzai Valley.

AT 1 a.m. on April 6th, 1891, a telegram was received from the Quartermaster-General at Simla, directing the Regiment to march at once, maximum strength available, for Kohat, and later on the same day a telegram was received from the Assistant Adjutant-General, Punjab Frontier Force, dated Reserve Column, Darband, Black Mountain Field Force, to proceed by double marches to Kohat.

The Regiment marched, 428 rifles strong, at 1 a.m., April 7th, reaching Banda, distance fifty-six miles, at 9.20 a.m. on the 8th. From Banda they marched leisurely to Kohat, reaching there on the 10th at 8 a.m. The entire distance to Kohat by the new road—namely, eighty miles—was covered, including halts, in seventy-nine hours, and the Corps was fit on arrival to continue the line of march, which it did at 1.15 the following morning, with No. 3 Mountain Battery (European), reaching Chili Bagh at 5.30 a.m. (eleven miles), where a halt of six hours was made. The march was resumed at 1 p.m., and Hangu, twenty-six miles from Kohat, was reached at 7 p.m. Everyone was heavily drenched at the nineteenth milestone by a thunderstorm. The men of the Regiment passed the night as they were in Masjids, the Dak Bungalow, and out-houses, some being very kindly accommodated in the camp by a detachment of the 19th Bengal Lancers. On the evening of April 12th the Regiment moved into bivouac with the 15th Sikhs.

A congratulatory telegram was received from the Commander-in-Chief, expressing " much pleasure at the rapid march of the 6th Punjab Infantry from Bannu."

The weather continued to be very wet, and the bivouac of the 15th Sikhs and the 6th Punjab Infantry being in ploughed land, the health of the Regiment was well and successfully tested.

On April 16th the Regiment marched, together with the 15th Sikhs and the 19th Punjab Infantry, under Lieut.-Colonel Sandilands, to Darband, and there bivouacked. On April 17th, No. 3 Column, Miranzai Field Force, consisting of three guns, No. 3 Mountain Battery, the 6th, 19th and 29th Punjab Infantry, was formed under the command of Lieut.-Colonel Brownlow, and after acting as support during the advance of No. 2 Column, proceeded by the new military road by the village of Dhar to Sangar. This proved to be a fatiguing march. The slight opposition to the advance of No. 3 Column was met by the 29th Punjab Infantry, which Regiment was allowed to lead on this occasion, on account of the losses it had sustained on April 4th.

On the 18th the advance on Surtoh was made, the 3rd Column leading, with the 6th Punjab Infantry at the head, followed by the 19th and 29th Punjab Infantry. The advance of the Column was covered by the fire of No. 3 Mountain Battery Royal Artillery. In the advance the Regiment lost one naik killed, one sepoy severely and three sepoys slightly wounded. After the clearance of Surtoh the advance continued, and the Column halted at Mastan, covered by the Regiment, which occupied the furthest hamlet of Sari Garhi, with three companies as piquets on the hill facing Gulistan. The Regiment was recalled in the evening and bivouacked with the column at Mastan, which is where Fort Lockhart is now situated.

On the 19th, officers and men all received a blanket and a great coat from their field kits, which had been left behind at Sangar. The enemy, who occupied in force the village of Gustang, kept up a lively fire on the ridge extending from the Crag Piquet (which is where the present butts are situated), held by the 19th Punjab Infantry, down over the water nullah, which was protected by the Regiment. At this period Subadar Mowaz Khan was slightly wounded. Later on the Regiment returned to the Column bivouac, and shouting, yelling and firing from all sides was kept up by the enemy till midnight.

On the 20th orders were received to proceed to Surtoh, which was held by two companies of the 2/5th Gurkhas. At 1 p.m. the Regiment was recalled to Mastan, where an hour's halt was given to allow the men to rest and feed before proceeding with three guns, No. 3 Mountain Battery, and the 19th Punjab Infantry, under the command of Lieut.-

Colonel Sandilands, to attack the village of Gustang, simultaneously with the attack on the villages of Sari Garhi by the remainder of the Column, reinforced by the 60th Rifles, 2nd Punjab Infantry, and 1/5th Gurkhas. The advance on Gustang was unopposed and the village was destroyed.

On the 22nd the Regiment, as the advanced guard of a force under Lieut.-Colonel Reid, occupied, after some resistance, the elevated village and tower of Bazaie. On this occasion the capture of the height was opportunely assisted by an excellent shot from No. 3 Mountain Battery, commanded by Lieutenant Parker, R.A. The various hamlets of Bazaie were then destroyed, and the enemy were seen flying across the Khanki River in numbers.

On the 24th the Regiment proceeded as part of a force under Lieut.-Colonel Sandilands to occupy the villages of Jandalai, Dupkai, Udahke and Kazakou. These villages were taken without any resistance being offered, and the towers were mined for blowing up and the houses were prepared for burning, when orders were received to stop all punitive measures and retire.

The Regiment remained in camp at Mastan until May 6th, tents and field kits having been got up in the meantime, and on the 7th proceeded to Gulistan to join the 2nd Column, under Lieut.-Colonel Turner.

They remained in bivouac at Gulistan on the 8th and 9th, and on the 10th proceeded to Krappa as part of the reinforced 2nd Column. On the 12th the Column moved on to Sardari, and on the 13th to Khanki Bazaar. On the 14th it moved direct back to Krappa, and on the following day to Gulistan. The Regiment returned to Mastan on May 16th, the Column having experienced no opposition during its march down the Khanki Valley. The 1/5th Gurkhas from the 1st Column at Sangar occupied the tents of the Regiment which were left standing at Mastan during its absence with the 2nd Column.

During the operations, and the subsequent occupation of the Samana Range, every variety of weather was experienced. Lieut.-Colonel A. N. Sandilands and Surgeon-Major J. T. B. Bookey, I.M.S., were mentioned in General Sir William Lockhart's despatch.

On June 8th the Regiment commenced its return march to Bannu via Togh and the Shakkar Khel Valley, arriving in quarters on the morning of the 18th.

During the expedition the Regiment was filled up to the total fighting strength of 712.

On November 5th, 1891, the Regiment was inspected by His Excellency the Commander-in-Chief, who was pleased to express satisfaction with the Corps generally, and desired his appreciation of the rapid march made by the Regiment to Kohat to join the Miranzai Expedition to be made known to all ranks.

On December 1st the Regiment was equipped with the Martini Rifle, Mark II.

1892. On March 2nd, 1892, Lieut.-Colonel Sandilands vacated command of the Regiment on attaining fifty-two years of age, and Major J. E. Mein was appointed Commandant from the same date.

A detachment, strength 2 British officers, 3 native officers, 16 non-commissioned officers and 134 sepoys, left Edwardesabad for Jandola on August 22nd, under the command of Captain D. J. O. Taylor.

The Regiment marched from Edwardesabad on November 9th, and relieved the 3rd Sikh Infantry at Kohat on the 14th of the same month.

1893. No musketry was carried out during the year 1892–93 owing to heavy duties and detachments that had to be provided by the Regiment while they were stationed at Kohat.

The detachments from Jandola and Edwardesabad, under Captain Taylor, rejoined Headquarters on March 19th.

On April 1st telegraphic instructions were received for the Regiment to proceed to the Kurram Valley, to relieve the 2nd Punjab Infantry, all recruits to accompany the Regiment, and no depot to be formed at Kohat. On April 9th the Right Wing, under the command of Captain E. W. Cunliffe, marched from Kohat *en route* to Kurram, and on April 25th, on arrival of a wing of the 2nd Punjab Infantry at Kohat, the Left Wing and Headquarters marched, reaching Camp Sadda on May 3rd.

From May 3rd to the 15th the Regiment was split up into detachments at Fort Kurram and Kulachi, and on May 16th moved into a summer standing camp on a spur of a hill running down from a high range on the north of the valley, near the village of Malana, the elevation of the camp being about 6,000 feet. The Regiment remained here until October 25th, together with two guns of No. 3 Peshawar Mountain Battery and one squadron 5th Punjab Cavalry. The force was then ordered to proceed down the valley to a more sheltered spot for the winter.

The Regiment was relieved by the 21st P.I. on November 29th, and commenced its return march to Kohat on the following day, arriving there on December 6th. The health of the troops throughout their stay in the Kurram was very good.

On July 20th Lieutenant W. Seton Browne died of enteric fever at Camp Malana after four days' illness. He was succeeded as Adjutant by Lieutenant T. C. Plowden, who was appointed on August 19th.

Subadar-Major Mowaz Khan, who had succeeded Subadar-Major Bhola Singh, retired to pension on December 31st after thirty-two years' service in the Regiment.

The Regiment was stationed at Kohat throughout the year 1894 until December 4th, when it proceeded to Bannu to join the Waziristan Field Force.

1894-95, Operations in Waziristan.

On March 5th and the following days the Regiment was inspected by Major-General A. P. Palmer, C.B., who gave a most satisfactory report as the result of his inspection.

Surgeon-Lieut.-Colonel J. T. B. Bookey, who had been associated with the Regiment for nearly twenty years, reported his departure on November 13th to take over medical charge of the 5th Gurkhas at Abbottabad.

On November 29th telegraphic instructions were received warning the Regiment to prepare for active service in Waziristan. On December 3rd orders were received for the march to Bannu, and early next morning the Regiment left Kohat, strength 6 British officers—Major Mein, Captains Cunliffe, Taylor and Hutchinson, and Lieutenants Limond and Browne—and 672 rank and file, reaching Bannu on the 8th of the month. On the 10th Lieutenant H. W. R. Senior, 44th Gurkha Rifles, and Lieutenant F. B. Hill, 34th Pioneers, reported their arrival as attached officers.

On December 11th the Bannu Column, composed of the following troops, was assembled at Mirian: 3rd Punjab Cavalry, No. 1 Kohat Mountain Battery, 1st Sikhs, 2nd Punjab Infantry, and the 6th Punjab Infantry; the whole being under the command of Colonel Egerton, D.S.O.

On December 17th the advance across the border was commenced via the Khushurah Nullah, with Makin as the objective, which place was reached on December 22nd without any opposition being offered *en route*, with the exception of some sniping at the baggage and rear-guard when ascending the Razmak Kotal. The camp, however, was fired into each night, but no casualties occurred.

A halt was made at Makin, where the 2nd Brigade, under Brigadier-General Symons, together with the Headquarters staff, had arrived, until Christmas Day, when four flying columns were despatched with three days' provisions and no tents, to scour the valleys round Pir Ghul. The Regiment formed a portion of a column under Major Mein, that proceeded up the Darra Valley to Mandesh and Razin, returning to Makin on the night of December 27th.

Captain Hutchinson left Makin on December 31st to take over command of the Depot at Kohat.

On December 31st the Bannu Column, with the addition of No. 1 Company Bengal Sappers and Miners, equipped entirely with mule transport, and rationed for ten days, started to scour the country round the Shabutu, and Shuza Nullahs, where it was reported that the Mahsuds were grazing their flocks. During this tour the troops had a lot of hard and difficult marching, two steep passes having to be negotiated, and the only supply of drinking water being brackish. All defences of the enemy were thoroughly destroyed *en route*, as on previous occasions.

On January 8th, 1895, the column surprised at dawn a big gathering of Mahsuds, and captured 3,000 head of cattle, which were driven into Jandola the same day, over some twenty-four miles of rugged country. On January 7th, Subadar Sone Khan, "H" Company, died from the effects of an attack of pneumonia, contracted on the march. His corpse was carried into Jandola, and buried there.

At Jandola a concentration of the three brigades employed in the expedition under Lieut.-General Sir W. S. A. Lockhart, K.C.B., C.S.I., took place. On January 12th the brigades were put in motion again, the Bannu Brigade receiving orders to form a flying column, consisting of 400 rifles from each regiment, and four guns to proceed to scour the country round Babu Garh, and thence to Jani Khel, where they were to concentrate and await further orders. The remainder of the column, with all sick and camel transport, proceeded on the same date to Jani Khel via Tank, and the Bain Pass. They reached Jani Khel on January 16th, where they awaited the flying column, which arrived on the 19th. The whole column then marched on the following day to Barun, a few miles out of Bannu, where they remained encamped until February 8th, when the camp site was changed to a spot near the Tochi Post. The force remained here until February 22nd, when an advance was made up the Tochi Valley.

As an expedition, the 3rd Mahsud Waziri was a disappointing one to the soldier, as no opposition whatever was offered by the enemy to the troops at any one point. Although there was no fighting, the men had to endure great hardships from the climate and a good deal of heavy fatigues, rearguards, and night piquets. The marches were trying owing to the beds of streams having to be used as roads, which necessitated constant wading through water. A lot of snow fell during December, and on some nights about twenty degrees of frost was experienced. However, all was cheerfully borne and endured by the men, and with the exception of pneumonia, of which there were twenty-one fatal cases, the health of the Regiment throughout the expedition was good.

On February 22nd the Bannu column together with the following Divisional troops—Border Regiment, 20th P.I., 1/5th Gurkhas, and one company Bengal Sappers and Miners as Delimitation Escort, the whole under the command of General Sir W. Lockhart, commenced the march up the Tochi Valley, which had never previously been entered by any British troops. Lieutenant A. S. Stephen, 36th Sikhs, joined the Regiment as an attached officer on February 26th.

On the march up the valley, the Regiment was detailed to guard the line of communications, furnishing posts at Shinkai Kotal, Idak, and Mohamed Khel. Two hundred and fifty rifles, with Headquarters, proceeded to Sherrani, which was reached on March 2nd. On March

10th Headquarters left Sherrani to patrol the valley, Major Mein being placed in charge of the communications, while the remainder of the column proceeded to Dwatoi, as a support to the Delimitation Party, who now commenced their work.

On March 8th Sir W. Lockhart with staff and divisional troops left Sherrani for the return march to India. On March 22nd Colonel Meiklejohn, C.M.G., 20th P.I., was appointed to command the troops in the Tochi, *vice* Colonel C. E. Egerton, D.S.O., relieved.

On March 31st the column concentrated at Miran Shah, and the Delimitation Party rejoined on April 5th, having completed their work. It was now thought that a retrograde movement would be made out of the valley, but with the exception of the 1st Sikhs, who were ordered back to Bannu, the force received instructions to stand fast. Strong posts were now commenced to be built at Saidgi, Idak, and Miran Shah. The 2nd Punjab Infantry was employed in garrisoning them. The remainder of the force left Miran Shah on May 13th *en route* to Dehgan, where it was decided to form a standing camp for the summer.

On this date a most deplorable incident took place. Lieutenant A. Limond, Officiating Adjutant, while walking in Camp Boya, escorted by his orderly, was surprised and attacked in a most dastardly manner by four Dauri fanatics, and sustained a severe stab in his abdomen, from the effects of which he expired on the following day, May 14th, and thereby the Regiment lost a most valuable and esteemed officer, and one of great promise. It is satisfactory though to place on record that none of his murderers survived to brag of their dastardly deed, but all were despatched, and met their end at the hands of men of the Regiment.

On May 16th Dehgan was reached, where a strong entrenched camp was thrown up. Captain E. W. Cunliffe proceeded from Dehgan on June 7th to relieve Captain J. W. C. Hutchinson at the Depot.

The Regiment remained at Dehgan until July 20th, when it was ordered to occupy posts at Miran Shah, Idak, and Saidgi, for a month's duty on the line of communications.

The force concentrated at Miran Shah on August 25th, on which date the Regiment was relieved of its duties on the line of communications, and joined the camp. Lieutenant Fenner relieved Major Cunliffe, who rejoined from the Depot on October 11th.

The Regiment again performed a month's duty on the line of communications from October 24th. The Depot moved from Kohat to Bannu on December 5th.

1896. The Regiment was inspected at Miran Shah on January 20th, 1896, by the Officer Commanding Troops in the Tochi, and again took the line of communication posts from January 24th.

On February 8th the Tochi Valley troops were reduced by a regiment of infantry, and the 20th Punjab Infantry left for Peshawar.

Major-General Sir Power Palmer, K.C.B., Commanding the Punjab Frontier Force, visited Miran Shah on February 17th, and at the Brigade parade of the Tochi Valley troops, on February 18th, presented the Indian Order of Merit (3rd Class) to No. 2960 Lance-Naik Bal Singh, "A" Company, awarded by the G.G.O. No. 63, dated January 17th, 1896, for "conspicuous gallantry in the Tochi Valley on May 13th, when he, while acting as orderly to Lieutenant Limond, rushed to the rescue of that officer, who was attacked by four fanatics. He dispersed the four men, and engaged one in single combat, wresting from his hand the knife with which Lieutenant Limond had been stabbed, and receiving in the mêlée a bad cut on his hand, and a bullet wound in his side."

On March 13th Mr. Casson, Political Officer, Tochi Valley, was stabbed by a fanatic, who was most gallantly seized by Jemadar Zaffar Khan, of the Border Police, and killed by Captain Eales, 2nd Punjab Infantry, then commanding Idak Post. Two men were also taken prisoner.

On the following day Major E. W. Cunliffe assumed command of the Tochi Valley troops. The command was made a second-class station with a second-class station staff officer from April 1st.

Preparatory to the redistribution of the troops of the Tochi Force, the leave and furlough men of the Regiment were recalled, and on August 30th, 1896, the Regiment marched to Datta Khel, together with a detachment of the 1st Sikhs and four guns, a post being established at Boya on the way up, and held by a company of the 1st Sikhs.

D 2

An entrenched camp was made at Datta Khel above some water springs, 1¼ miles to the south-west of Datta Khel village, in the Tsirai Plain at an altitude of about 4,300 feet. The maxim gun attached to the Regiment was dragged up the bed of the Tochi River to Boya and on over the Tut Narai Kotal by two mules without damage. The inhabitants of the plain (Darwesh Khel Wazirs) showed a passive acquiescence to the arrival of the troops, but supplies were not readily brought in, milk and eggs being especially difficult to obtain. Soon after the arrival of the troops, thieves fired into the camp on two occasions but without result. The climate was very pleasant, except for some dust storms during the day.

On January 7th the Depot moved from Bannu to Dera Ismail
Khan, arriving there on the 12th. There they were
1897. joined by the Regiment, which marched from Datta
Khel on January 15th, reaching Dera Ismail Khan
on January 25th, after a period of over two years of camp life across the border.

On February 9th, and on the following days, the Regiment was inspected by Major-General Sir Power Palmer, K.C.B.

Owing to the stoppage of leave and furlough during the two previous years while the Regiment was in the Tochi Valley, a special concession was granted this year by allowing double the regulated number of men to proceed on furlough, which opened in February, after the inspection. The first batch enjoyed their full furlough, but the second batch, who left in July, were recalled a fortnight after departure owing to disturbances that broke out on the North-West Frontier towards the end of July. All British officers on leave were likewise recalled. In this month the Regiment received a fresh issue of Martini-Henry rifles, Mark II.

On September 22nd Major Cunliffe proceeded to join the 3rd Sikhs in the Tirah Expedition, and rejoined again at Dera Ismail Khan on October 31st.

On October 20th the Left Wing under Major MacMullen left for Dera Ghazi Khan to relieve the 2nd Sikhs, who were ordered to Kohat for garrison duty. On November 2nd telegraphic instructions were received for headquarters of the Regiment to be transferred to Dera Ghazi Khan, and for the Right Wing to relieve the 2nd Sikhs in outpost duty at Kajuri Kach, Nili Kach, Jatta, Draband, and Tank. Headquarters arrived at Dera Ghazi Khan on November 14th to relieve the above-named posts. The Regiment, after a very brief spell of cantonment life, was again split up into numerous detachments.

On December 23rd a party of 97 rank and file, under Subadar Ditta Khan and Jemadar Wali Khan, left by rail *en route* to Peshawar, to join the 2nd Punjab Infantry in the Bara Valley, with the Tirah Field Force.

During the year 1898, the Regiment remained at Dera Ghazi Khan, providing detachments as before. The 100 rifles sent to Tirah Field Force remained with the 2nd Punjab Infantry.

The Regiment remained at Dera Ghazi Khan, still providing the same outposts during the year 1899, until the Gumal outposts were relieved by the 27th Punjab Infantry towards the end of November, and after concentration at Tank, marched to Kohat, under command of Captain Fenner. Headquarters at Dera Ghazi Khan were relieved by a wing of the 40th Pathans on December 7th, and arrived at Kohat on January 5th the following year.

1900, Jubilee Celebrations.

The following is an extract from the remarks made by General Sir A. P. Palmer, K.C.B., on the occasion of his reviewing the troops at Kohat at the Jubilee Celebrations of the Punjab Frontier Force, on March 10th :—

1st P.C., 3rd P.C., No. 4 H.M.B., Punjab Garrison Battery, 1st Sikhs, 4th P.I., 6th P.I.

" It has given me great pleasure as an old Commandant of the Punjab Frontier Force to review to-day the troops at Kohat, and I only regret that it has not been possible to get together more of the regiments and batteries on this historic occasion, but owing to the nature of their duties, it is, of course, impossible for many to be collected in one place, even to celebrate their jubilee.

" These duties, which necessitate the greater part of the force being split up into small detachments guarding the North-West Frontier of the Indian Empire, whilst affording many opportunities for learning some of the more important duties of soldiering, leave little time for the practice of ceremonial parades, so that it is all the more creditable that you should be able to go through parade movements in a manner that would not disgrace the regiments of our Army, who spend the greater part of their service in comfortable cantonments. No reviewing officer could help being struck with the workmanlike and soldierlike bearing of the men whom I have seen on parade to-day, and I am proud to remember that I was once at their head, and to know that since I have been associated with the Punjab Frontier Force, it has in no way lost its efficiency; but while the same fine spirit, for which it has always been famous, remains as high as ever, it has come on in keeping level with the times, and its regiments are fit to be brigaded with any troops in our Army, either in peace or war. I congratulate the Frontier Force on the name it has gained, not only for efficiency, but for loyal service, especially when loyal service was so valuable to the British Empire in India forty-three years ago.

"As 'Warden of the Marches,' the Punjab Frontier Regiments have been making history for half a century, and have earned the gratitude of the rest of the Army in India, who have been able to rest secure like troops in camp, while the Punjab Frontier Force have kept watch and ward for fifty years of outpost duty."

The above extract is taken from Punjab Frontier Force Order No. 160, dated March 21st, 1900.

His Excellency The Viceroy, Lord Curzon of Kedleston, visited Kohat on April 23rd and 24th, and the Regiment furnished his guard of honour at the Mess on the 23rd, and 100 rank and file under two British officers piqueted the heights between Gumbat and Tilkham on the following day, the date of his departure.

On July 4th a short but severe bout of cholera broke out amongst the whole of the troops in the garrison, who were moved out into camp. Station duties meanwhile had to be carried on. The Regiment returned to barracks on July 25th, after being ten days clear from the last case. There were 37 seizures and 15 deaths in the Regiment.

On September 26th the Regiment proceeded to Bannu, for garrison duty there, while the Bannu moveable column proceeded to the Tochi Valley to punish the Maddu Khels. The column returned on October 24th on completion of its duty, and the Regiment left for Kohat on the following day.

BRITISH AND INDIAN OFFICERS AT KOHAT, 1901.
Colonel J. E. Mein seated in the centre.

At the end of 1900, the institution of double companies, instead of wings was brought into force, and the companies of the Regiment were accordingly linked as follows :—

1st Double Company	"A" Sikhs "G" Sikhs
2nd Double Company	"B" Yusafzais "D" Bangash
3rd Double Company	"C" Dogras "E" Dogras
4th Double Company	"F" Khattaks "H" Punjabi Mohammedans

1901. Colonel J. E. Mein relinquished command of the Regiment on March 3rd, 1901, on completion of two years' extension of command. The following Regimental Order was issued by Colonel Mein on March 2nd :—

"On relinquishing command of the Regiment this day, Colonel Mein bids farewell to all ranks, and thanks the British and native officers for the cordial support and assistance they have rendered to him during the nine years he has had the honour to command the Regiment. He also tends his thanks to the rank and file for their orderly conduct and good behaviour during these past years, thereby upholding the credit of the Regiment, and he trusts they will continue to keep up the high reputation established, and finally wishes to one and all, good luck and success."

Major E. W. Cunliffe succeeded Colonel Mein as Commandant.

During August the Regiment was rearmed with the Lee-Metford rifle Mark I, and on the 23rd of that month, the Regiment marched from Kohat in relief of the 1/5th Gurkha Rifles on the Samana.

On October 2nd the Ali Sherzais and the Mahmozais came down and burnt the Alizai village of Zir Girri, near the spot of Shinawari, of which Subadar Baz Khan, 6th Punjab Infantry, was in command. He turned out at night, attacked and beat off the raiders with great promptitude, and so saved the village of Shinawari, and the Alizai villages from being burnt. Subadar Baz Khan received the commendation of the Lieutenant-General Commanding the Forces, Punjab, for his "prompt action in attacking and beating off the raiders."

1902. In June, 1902, the posts of Shinawari and Fort Cavagnari were taken over by the Samana Rifles (present Frontier Constabulary), and our men were withdrawn, one company to Fort Lockhart, and two companies to Thal, which was taken over from a detachment of the 19th Punjab Infantry by Captain Fenner.

An Adam Khel Afridi thief was shot dead by a patrol of the Regiment while he was prowling round the transport mule lines at night, on September 6th.

The Regiment was not relieved on the Samana this winter, and so endured another cold weather on outpost duty. The Sikh companies at Thal had suffered from malaria and bowel complaints, so that when they returned to Fort Lockhart, their admissions to hospital during October and November rose as high as 69 in the double company.

The Sikh double company was detailed as escort to the British Commissioner to delimitate the boundary between Afghan and Turi territory, and proceeded to the Kurram Valley on February 22nd, 1903.

Lord Kitchener, Commander-in-Chief, and his staff visited the Samana from Kohat, returning on the same day, January 21st, 1903.

On May 11th, 1903, the Regiment was relieved on the Samana and outposts by the 1st Sikhs, and reached Kohat on May 14th, after having been on the Samana for nineteen months. The detachment from Thal rejoined at Kohat on May 18th.

During 1902 and 1903 the following officers were attached to the 2nd Sikhs for service in Somaliland:—Captain P. B. Forster, Lieutenant A. M. Gillies, Captain F. D. Browne, and Captain C. C. Fenner. Five Indian officers were also attached to the 2nd Sikhs during these years.

By Indian Army Order 181 of 1903, the Regiment received the new title of—

"59th Scinde Rifles (Frontier Force)."

BREAST ORNAMENT (bronze).

POUCH ORNAMENT (silver).

BUTTON (silver).

SHOULDER NUMERAL (white metal).

CAP BADGE (silver).

CHAPTER V

THE 59TH SCINDE RIFLES (FRONTIER FORCE), 1904–1914.

ON June 8th, 1904, the Regiment marched from Kohat to Bannu, where they relieved the 54th Sikhs (Frontier Force).

Major-General B. R. Chambers was appointed Honorary Colonel of the Regiment in the London Gazette dated May 13th, 1904.

Lieut.-Colonel E. W. Cunliffe was promoted Brevet Colonel by Gazette of India 951 of 1904.

During the year Captain Kirkpatrick and Lieutenant W. Campbell were attached to other corps for service with the Tibet Mission, both rejoining before the end of the year.

At the beginning of 1905 Captain F. D. Browne was wounded on the golf links at Bannu by a Ghazi fanatic. He proceeded on eight months combined leave, and was subsequently transferred to the 56th Rifles (Frontier Force).

On November 18th, 1905, the Regiment marched for the Rawalpindi manœuvres, in which it took part, being in the 3rd (Frontier) Brigade, 1st Division, Northern Army. It also took part in the Royal Review before H.R.H. The Prince of Wales on December 8th, on which occasion some 55,000 troops were present. The Regiment returned to Bannu on December 24th.

The Regiment proceeded on a short trans-frontier reconnaissance on March 19th, 1906, but returned two days later on account of inclement weather. They remained at Bannu until November 13th, when they marched to Peshawar, arriving there on November 22nd.

During the year 1906 a party of 2 native officers and 14 rank and file were entered for the rifle championship at Meerut, and prizes were won to the value of Rs145 in the individual competitions and Rs147 in the team competitions.

On September 30th Lieutenant Gillies (Quartermaster) died from

ZAKKA KHEL EXPEDITION, 1908.

1.—1st Brigade Camp, Ali Masjid.
2.—Officers' Mess, at Chora. Captain J. Husband, Major C. C. Fenner, Lieut. R. D. Inskip, Lieut. A. H. Burn.
3.—Jirgah accepting terms at Walai.
4.—Group at Jamrud.

ZAKKA KHEL EXPEDITION, 1908.

1.—Trenches at Walai.
2.—Chora Fort.
3.—Cavalry during Retirement from Halwai.
4.—Officers present at Matta : S.M. Ditt Singh, Lieut. Scott, Lieut. Anderson, Col. Carruthers (nearest camera).

an accident at polo. He was a great loss to the Regiment, as he was of a most cheerful nature and was a most promising officer.

1907.

H.M. The Amir of Afghanistan visited India at the beginning of 1907, and the Regiment furnished a guard of honour composed of 2 native officers and 100 rank and file, under the command of Lieutenant Plowden, over the Amir's camp from January 3rd until the 6th.

During the month of February typhus fever was brought into the Regimental Hospital through some Mule Corps drivers who were attached to the hospital for treatment. Four non-commissioned officers and sepoys died of it, and many were attacked but recovered. The Left Wing only was attacked by the disease, and the whole wing was accordingly segregated in tents. With great care and excellent attention and arrangements on the part of the medical officer, Captain Husband, I.M.S., this disease was soon stamped out, though it continued with the Mule Corps.

Perbhat Chand, son of Subadar-Major Bidi Chand, Sirdar Bahadur A.D.C., was appointed as a direct Commissioned Jemadar on probation from March 20th. He is the second recorded case of a "direct commission" Indian officer in the Regiment, the first having been Jemadar Jehan Dad Khan, who was appointed as a Jemadar on probation on May 4th in the previous year. On November 26th, 1907, Jemadar Bishen Singh was also appointed as a jemadar on probation.

Subadar-Major Baz Khan, Bahadur, retired to pension. Some years later he received the title of Khan Sahib from the civil authorities in recognition of good work. This Indian officer now has a son in the Regiment, Jemadar Makhmad Yusaf.

1908. Zakka Khel Expedition.

On February 13th, 1908, the Regiment marched out of Peshawar to take part in the expedition against the Zakka Khel. They were brigaded with the 1st Brigade Bazaar Valley Field Force, the Brigade consisting of—1st Bn. The Royal Warwicks, 1/5th Gurkhas F.F., 53rd Sikhs F.F., 59th Scinde Rifles F.F., commanded by Brigadier-General C. A. Anderson, C.B.

After a halt at Jamrud, Ali Masjid was reached on February 14th. The next day two brigades entered the Bazaar Valley, the Regiment being ordered to halt at Ali Masjid that day, to help the 25th Punjabis to guard the camp and supplies.

On the 16th the Regiment marched to Chora without incident, escorting the Brigade Supply Column of 1,694 mules over the Chora Kotal on an excellent road, which had been improved on the south side by the Sappers and Miners. The 25th Punjabis piqueted the hills for two miles, and the 23rd Pioneers from Zera to Chora and towards Ali Masjid, the Regiment supplying the remaining piquets.

On arrival at Chora, Colonel Cunliffe took command on the 17th, on the departure of the G.O.C. 1st Brigade and Staff for Walai. Convoys of mules proceeded daily both ways, the Regiment supplying the piquets from Chora Kotal to Chora. Nothing of interest occurred except for the shooting by a 23rd Pioneer piquet of a shepherd, whom they had detained until the arrival of a Pushtu speaker. The man suddenly seized the havildar by the beard, struck at him with an axe, and then fled. As he refused to stop, he was promptly shot. On the 18th the piquets of the Regiment on the south side of the Kotal were fired on, but no damage was done.

On the 20th the Regiment moved up to Walai in relief of the 45th Sikhs. On the night previous the signallers working a lamp in camp were sniped.

On the following day the Regiment formed part of the 1st Brigade in the attack on Halwai.

The 1st Brigade moved over a Kotal south of Khwar, while the 2nd Brigade held the China Hill and advanced from the east against Halwai. The Regiment acted as rearguard during this movement, there being a few of the enemy on the hills north of Khwar. After crossing the Kotal the 1st Brigade made a frontal attack on Halwai, and the Regiment acted as right flank guard to keep off an attack which was expected from the direction of Nikai. On Halwai being taken and blown up, the retirement began, the enemy showing in some numbers both from the hills and before Halwai and Nikai. The Regiment acted as rearguard in conjunction with the 53rd Sikhs, finally retiring through them on nearing the China Hill. In the retirement there was only one casualty in the Regiment (one sepoy wounded).

Three companies were later employed in covering the retirement of the 28th Punjabis from China Hill.

On February 23rd the Regiment accompanied the 1st Brigade to complete the demolition and to bring away wood from China Hill. At 2.15 p.m. the order to retire was given, and the Regiment once more acted as rearguard, keeping in touch with the Warwicks, who were acting as left flank guard. The retirement was very feebly followed up, and at some distance.

On the following day the Regiment took over the piquet on 45th Sikh Hill, another on the north of the road to Chora, and also piquets round the camp. The remainder of the Regiment formed part of the escorts to convoys to and from Chora.

In the meantime negotiations were opened with the Afridi Jirgahs, and an armistice was declared.

On the 29th the Regiment formed part of the troops covering the retirement from Walai. There was no firing on the part of the enemy, and the Brigade reached Chora at 4 p.m.

On March 1st the Regiment piqueted the route during the retirement from Chora to Lalachina. They then proceeded to encamp at Ali Masjid to guard some 30,000 maunds of supplies, while the remainder of the troops marched to Jamrud.

The Regiment was congratulated on the way it had played its part during the expedition, especially in the action at Halwai, by the G.O.C. 1st Brigade, and also by Sir James Willcocks, commanding the Bazaar Valley Field Force.

On March 1st Colonel Cunliffe, having completed seven years' command of the Regiment, handed over to Major R. A. Carruthers.

The following officers accompanied the Regiment during the Expedition :—

Colonel E. W. Cunliffe, Commandant.
Major R. A. Carruthers, Second-in-Command.
Major C. C. Fenner.
Major E. Kirkpatrick.
Captain T. L. Leeds.
Captain K. D. Murray, Adjutant.
Captain J. Husband, I.M.S.

Lieutenant B. E. Anderson, Quartermaster.
Lieutenant H. F. D. Stirling.
Lieutenant A. H. Burn.
Lieutenant H. N. Lee.
Lieutenant R. D. Inskip.

Events leading up to The Mohmand Expedition.

During the early months of 1908 the Mohmands began to show considerable restlessness, and raids became so common that it was found necessary to strengthen the garrison near the frontier Michni, Shabkadr and Abazai. This was done by sending men from each of the Peshawar Regiments. The 59th were ordered to send men to Michni, and 100 men under Captain Leeds were sent there in the middle of April. The Mohmands assembled in large numbers in the Gundab, and behind the Ali Kandi Pass. A force was pushed hastily out from Peshawar on April 20th, reaching Shabkadr that night; 150 men of the Regiment, under Lieut.-Colonel Carruthers, Lieutenant Anderson and Lieutenant Scott, formed part of this force and acted as advanced guard. The day was very hot and the march was most trying, everybody being completely done up on reaching the Kabul River. The force was made up of fragments of all the units in Peshawar, Brigadier-General C. A. Anderson, C.B., commanding. The force halted that night at Shabkadr, and a night attack was freely prophesied, but nothing happened.

On the next afternoon the force moved out to Matta Mogul Khel, two and a half miles north of Shabkadr, on the Abazai Road. Here a perimeter camp had already been built, and the detachment dug itself in on the south-west corner. The enemy, who had already come into collision with our troops, were visible on the hillsides in large numbers. They had built up sangars and planted their standards. The number was estimated at about 10,000, although a Mohmand who was present at the fight at Matta three days later afterwards stated that there were 30,000 Afghans present as well! In any case, there was a large gathering, and the tribesmen seemed very full of fight, and commenced sniping the camp that night, but without effect.

On the 22nd a column consisting of 100 rifles Royal Warwickshire Regiment, 150 rifles 59th and 50 rifles 28th Punjabis, and one half-section

Sappers and Miners, proceeded to Garhi Sadr, one and a half miles along the road to Abazai. A camp was built at Garhi Sadr on a slight eminence running across, and closing, the main road and overlooking a small canal, emanating from the Swat River. The site was good and a very strong camp was made, the Sappers and Miners giving invaluable help in devising cover.

The camp at Matta had also been considerably strengthened, and the garrison consisted of detachments of the 5th Fusiliers, Royal Warwickshires, 57th Rifles F.F., 19th Lancers, 21st Cavalry F.F., and one battery R.F.A. It was practically certain that there would be a night attack, and all were ready. At about 9 p.m. firing began at Matta, and the garrison replied with such enthusiasm that the firing got out of hand, and considerable ammunition was expended. The noise was so great that the horses of the cavalry broke loose and stampeded, cleared the trenches to the north, and galloped north along the Garhi Sadr—Abazai road.

Meanwhile in Garhi Sadr things were busy also. As was afterwards discovered, the tribesmen had selected Garhi Sadr for their real attack, Matta being merely contained. At 9 p.m. firing began from the south, where there was a broad nullah running from the Ali Kandi Pass to the road. The enemy were along the edge of the nullah, and among the trees fringing the road. The fire grew heavy, and the whole of the south face of the camp was busily engaged. All at once there was a great rush down the road from the south, and it was thought that the enemy were making a record Ghazi rush on the camp, and all the men stood to the perimeter wall, prepared to receive it. Suddenly a horse's head came over the parapet, and it seemed to " rain horses." It was the stampede from Matta, and the maddened horses threw themselves over the defences, through the camp, and out again over the other side, where they disappeared down the Abazai Road. The firing was stopped, and developments were awaited. Nothing happened, however, and it transpired that the enemy, thinking the cavalry were upon them, had fled in all directions and dispersed for the night. The stampede had taken the officers' chargers with it, and had knocked the barbed wire defences to pieces. In the morning a number of horses were recovered, including those of Colonel Carruthers and Lieutenant Anderson, but four horses lay dead in front of the trenches, and many more must have been badly hit. The 19th Lancers lost thirty-four horses over the incident.

On the following morning, the 57th Rifles relieved the detachment at Garhi Sadr, which returned to Matta Camp, where the Regiment held an outwork, a portion of the village at the north end. Matta was more or less left alone that night, but Garhi Sadr was attacked in force, and had a very exciting night. The officers of the 57th were full of praise of the trenches which our men had prepared for them, and attributed their small loss (one man killed) to the excellence of these trenches.

During the 22nd and 23rd General Anderson had been organizing his force, and collecting his fragments into regiments and corps units, and his preparations were now completed. He therefore decided to take the offensive, and deliver a blow at the enemy. The force at his disposal was distributed along the road as follows:—

Abazai—53rd Sikhs, F.F.	250 men
Garhi Sadr—57th Rifles, F.F.	200 men
Matta—19th Lancers	1 Squadron
21st Cavalry, F.F.	1 Squadron
R.F.A.	1 Battery
5th Fusiliers	250 men
1st Royal Warwickshires	300 men
28th Punjabis	50 men
59th Rifles, F.F.	150 men

On the morning of April 24th General Anderson, leaving the fifty men of the 28th Punjabis to hold Matta Camp, moved along the road to Garhi Sadr with his whole force. At a point half-way to Garhi Sadr, the 53rd and 57th detachments were met. The whole force halted, and turned to its left, and then advanced towards the hills to the west. The enemy's flags and sangars covered the face of the hills, and they were obviously there in large numbers. The infantry "brigade" consisted of only 1,110 rifles, in addition to two weak squadrons, and one battery. The 53rd Sikhs (250 rifles) were put in reserve, and the rest of the infantry in the front line. General Anderson's orders were for the force to advance as far as the first lot of underfeatures, but no farther. On the enemy's right a small green-topped hill (marked *A* in sketch) seemed to

Action at Matta.

SKETCH OF GROUND OVER WHICH THE ACTION OF MATTA WAS FOUGHT, 24TH APRIL 1908.

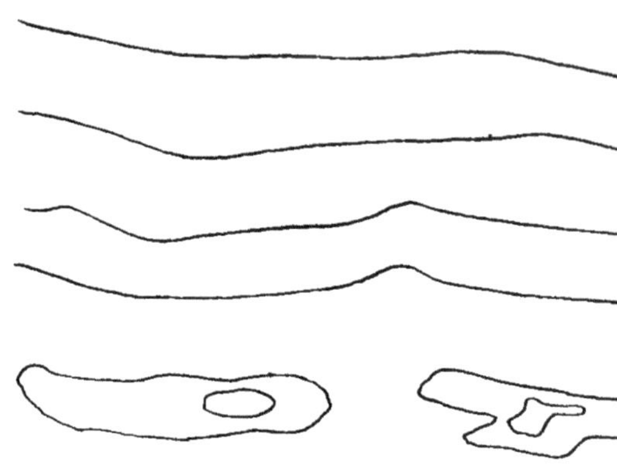

Open Plain

4 Mile

From Michni

Shabkadr

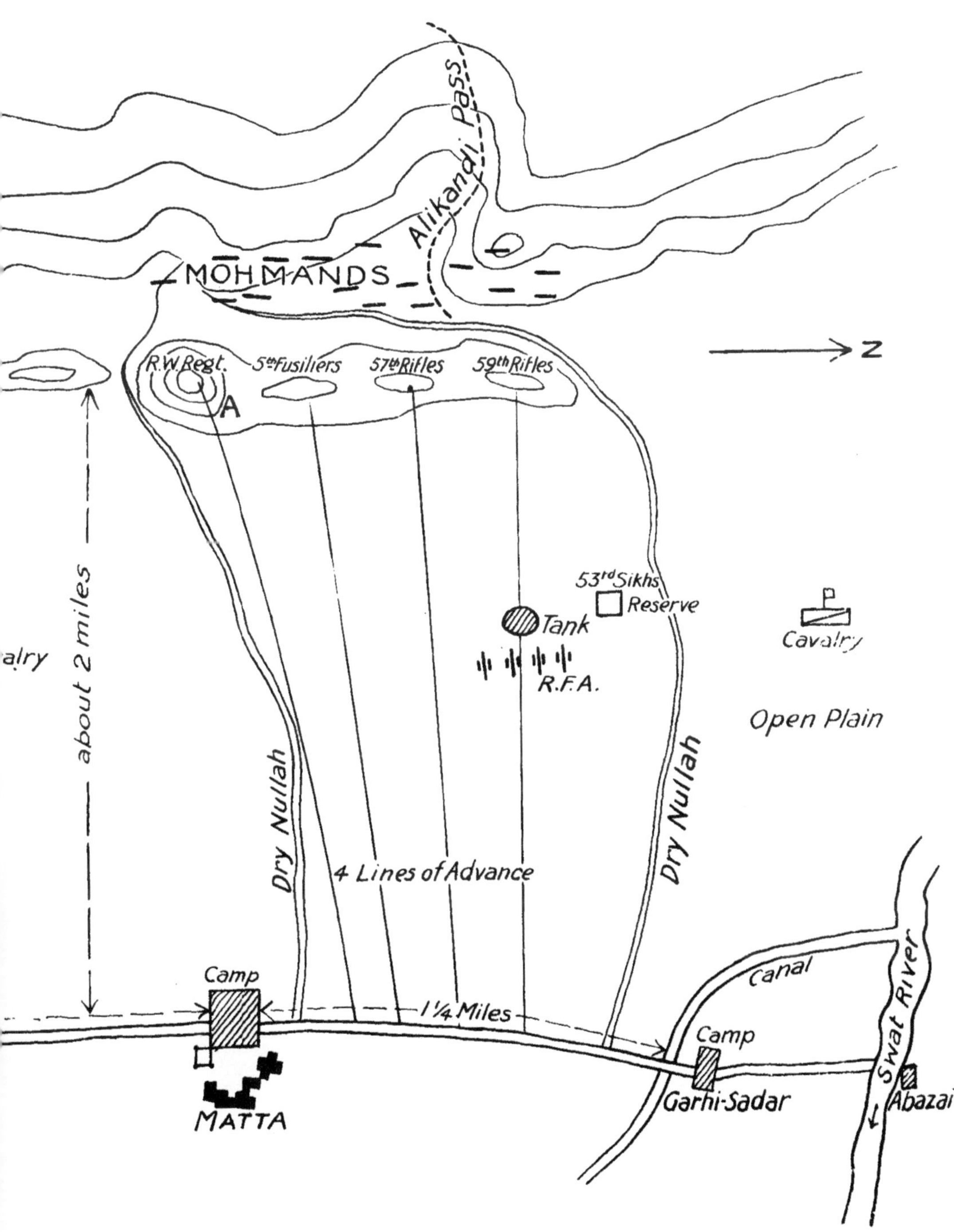

Alikandi Pass
MOHMANDS
R.W.Regt.
5th Fusiliers
57th Rifles
59th Rifles
A
Z
about 2 miles
alry
53rd Sikhs
Reserve
Tank
Cavalry
R.F.A.
Open Plain
Dry Nullah
Dry Nullah
4 Lines of Advance
Canal
Swat River
Camp
1¼ Miles
Camp
MATTA
Garhi-Sadar
Abazai
6/13TH F.F.R.

dominate the situation, and the Royal Warwickshire Regiment under Major Westmoreland was ordered to attack and capture it. On the right of the Warwicks were the 5th Fusiliers, and next to them the 57th, with the 59th on the right of the line. Lieut.-Colonel Carruthers commanded the infantry. The battery followed the infantry, and the squadrons guarded either flank. The reserve followed in rear of the right flank of the infantry.

The advance had not got very far before the enemy opened fire, which grew hotter as the hills were approached. The Warwicks on the left pushed straight on to their objective, the green hill, pushed parties up the slopes, collected men near the crest in some dead ground and finally rushed the summit with fixed bayonets, in fine style. The 5th Fusiliers, keeping level with the Warwicks, who set the pace for the whole line, won their point without much opposition.

The heaviest fighting of all fell on the 57th and 59th. In front of the right of the attack, the ground was broken into a series of ridges, immediately facing the Ali Kandi Pass, and here the Sangars were strongest, and the opposition most fierce. Bayonets were used, and sangar after sangar was only vacated when the bayonet came into play. The tribesmen fought very bravely, and only fell back towards the upper hills after taking and giving a lot of punishment.

The objective, the first line of underhills, having been gained, the force halted and was then ordered to retire. The retirement was carried out from the right. The 57th and 59th retired towards Matta Camp, inclining to the left behind the Fusiliers, who in turn on retiring, inclined behind the Warwicks. The green hill was the last point evacuated, but the retirement was very smartly carried out, and only the last party to leave the hill suffered any losses. A nullah runs round the southern end of the hill and as Lieutenant Martin* and the Royal Warwick scouts cleared the base of the hill, a party of tribesmen succeeded in getting round the base of the hill, and coming to close quarters, mortally wounded Lieutenant Martin. Under cover of the fire of the reserve, and the guns, the whole force then retired towards Matta Camp. There was no further attempt on the part of the enemy to follow up the retirement partly due no doubt to the presence of the

* Brother of Captain H. W. Martin, 59th Rifles (Frontier Force).

cavalry on the flanks, and also to the severe punishment the enemy had received. By noon the force was back in camp. Our detachment had one native officer wounded (Jemadar Jehan Dad), and 10 non-commissioned officers and men wounded, of whom one died afterwards.

This action at Matta was rather lost sight of in the larger expedition into Mohmand Land which followed it, but it was a noteworthy affair, and was regarded by those who took part in it, and in the subsequent expedition as the most important as well as the hardest fought action of the year. It had great political effects. Up to the fight, the tribesmen had been present in great numbers, and full of fight, but after Matta the enemy fled to their homes, thoroughly beaten. The tribesmen themselves admitted that they had suffered a defeat, and that their losses were heavy. A letter was picked up during the expedition three weeks later, in which the writer, addressing a big mullah, referred to "our defeat." The action was fought like a field day, an advance on to a position, carried out in accordance with a clearly defined plan, while the retirement was carried out in succession from a flank, under cover of the guns and reserve, with cavalry on the flank to check any pursuit.

Subadar-Major Ditt Singh distinguished himself greatly throughout the day, by his personal courage, and his leading of the charges on the sangars elicited unstinted praise from officers and men of the 57th Rifles, under whose observation he repeatedly came. He owed his Order of British India to his work on this occasion.

After the fight at Matta, the enemy dispersed, and hostilities ceased, but an expedition into the Mohmand country was decided on, and mobilization on a large scale ensued. The camps at Matta and Garhi Sadr were broken up, and a force of two brigades was in course of concentration at Shabkadr, when a rumoured attack on the Khyber Pass drew Sir James Willcocks and a mixed brigade in that direction. They met with little opposition, and the force which threatened the Khyber was in all probability the rumoured lashkar of 30,000 men returning to Afghanistan. They were probably "saving their faces" by talking of a big attack on the Khyber Pass, which, in the face of Sir James Willcocks' prompt action, they had not the courage to carry out.

The Regiment received orders to mobilize for the Mohmand Field Force on April 24th, 1908, and by May 12th everything was ready for the move forward. The Regiment was again in the 1st Brigade, in which the 22nd Punjabis had taken the place of the 5th Fusiliers, who were unable to join the expedition owing to an outbreak of cholera.

The Mohmand Expedition.

The big combined camp at Shabkadr was broken up on May 13th, and the 1st Brigade moved on to Dand, and thence to Ghalannai over the Kharrapa Pass on the following day. The road used was the same one that had been constructed in the 1897 campaign, but it had by this time practically disappeared. The pass was very rough, and the transport did not get into camp until 3 a.m. the next morning. On the 15th Nahakki was reached, a very dirty village with a poor water supply. The Nahakki Pass was not so difficult as the Kharrapa Pass, but gave considerable trouble, and would have been very formidable if it had been held. The force was now well within the enemy's country, and there was considerable sniping at night.

On May 16th the 1st Brigade marched up the Bohai Dag Valley to the west, to Khan Beg Khor, destroying several villages towards the Khapak Pass. In the afternoon the Brigade retired to Kasai, where the 1st Brigade made its headquarters for some days. During the retirement the Regiment acted as rearguard, and had to fight a running engagement for nearly seven miles. In this, the Regiment had seven casualties, including Subadar Fatteh Singh, a very fine soldier, whose loss was deeply felt.* On arriving in camp, just before dusk, the 59th were ordered to piquet a hill nearly a mile away from camp, to the south. The hill was held by the enemy, and three of our scouts were wounded, one mortally. That night, all the piquets were heavily attacked, with the exception of our own. The troops had got in late, the ground was very rocky, and there was not time to build up good sangars. One of the 22nd Punjabi piquets was very heavily attacked and sent in a lamp message that they could not hold out much longer. The piquet was reinforced by a company under Major Climo, under novel conditions. It was a very wet and dark night, and Major Climo, recognizing the difficulties of the relief, took out all the Regimental

* This Indian officer has a son (Lance-Naik Bhola Singh) serving in the Regiment at the present time.

drummers and surnais, and made all the noise he could, his men shouting encouragement to the piquet. This was too much for the enemy, who fled before this noisy reinforcement, and Major Climo was able to relieve the piquet, strengthen the defences, and remove the wounded without any loss.

On May 17th the force halted at Kasai, and on the 18th the Regiment went to take over the Nahakki Camp, while the 2nd Brigade went up the Bohai Dag.

On May 20th the 1st Brigade began its march to the north, with the object of penetrating to the Ambahar. The Regiment guarded the flanks and piqueted the hills until the Brigade had passed through the Lakai Tangi, and then the whole Brigade advanced across the open valley towards Yakh Dand. The enemy were visible on the hills round the village, and it was evident that there would be a fight. The 53rd and 57th were in the front line, and the 59th were in reserve. The action began at about 4 p.m., the leading regiments advancing towards Yakh Dand. The ground was very difficult, a network of gigantic nullahs that made anything like concerted action impossible. The tribesmen fell back on Yakh Dand, losing heavily, but it was not the General's intention to go as far as Yakh Dand that night, and after driving the enemy as far as the main nullah a little south of the village, the retirement was ordered. The tribesmen followed up with great dash and spirit, and there was considerable mixed fighting in the nullahs. Meanwhile the 59th had been preparing the village of Umra Killa for defence, and the Sikh double company, under Major Kirkpatrick, had been sent to hold a tank, 300 yards north-east of the village. As the front-line troops fell back, the tribesmen tried to get round on the right flank. No. 2 Double Company was sent out to strengthen that flank, and came in for a good deal of hot fighting. As our troops closed in towards the village, the tank became an advanced salient, and its evacuation increasingly difficult, as it would have been at once seized by the enemy and would have afforded him excellent cover from which to fire. Major Kirkpatrick timed his retreat very well, getting his ammunition out on men's heads, and dribbling his men back under cover. This party of the Regiment was the last to gain the village. The enemy had evidently lost somewhat heavily, parties with lanterns looking for the wounded were seen during the night among the nullahs, and the camp was very little molested.

On May 21st the force marched to Habibzai, and encamped in the open valley. A heavy night attack was prophesied, and it was rumoured that a band of Mohmands had arrived who made a speciality of night attacks. The sniping was heavy, and there was a good deal of noise from the enemy during the night, but no serious attack took place.

On the following day the Regiment formed part of an advanced column sent to attack and destroy the village of Bagh. The force came upon the enemy at about four miles from camp near Sherdal where a ridge crosses the valley, and a tangi commences. The column could not reach Bagh, and presently was joined by the advanced guard of the main body of the Brigade. No. 4 Double Company carried the ridge, and was pushing forward towards the main position, when the "cease fire" sounded, and orders came up from Sir James Willcocks to stand fast, as a jirga anxious for peace had come in. The enemy, however, were not quite so particular, and Captain Martin and his men had great difficulty in regaining the cover of our side of the ridge. Captain O'Grady had meanwhile been sent out to try to gain a ridge commanding the Bagh Valley, but he had found that his double company was insufficient for the purpose, and after putting up a capital fight, in which he himself was hit, he had to withdraw and join the Regiment. An armistice of some sort was arranged with the jirga, and the troops marched back to camp without further opposition. A quiet night was hoped for, but at 9.30 p.m. the usual attack commenced, and the camp was heavily peppered, there being a lot of casualties amongst the mules. The enemy were very bold and noisy on this occasion, getting close up to the camp along a nullah.

On the morning of the 23rd the force moved westward towards the Ambahar, and thence to Turu in the afternoon. No opposition was encountered, and a peaceful night was spent. The weather was very hot, and the Force was now in a dry, waterless country, the only water obtainable being that in the village tanks. This was both scanty and foul, and it was wonderful that the health of the troops remained so good as it did.

On the following day Kargha was reached, and for the first time good water was obtained. The tribesmen had collected here in great numbers, and it was thought that they would put up a big fight. The guns opened fire on the sangars which had been prepared in the open valley, and the 53rd Sikhs advanced over very broken ground, covering the left front. A wing of the 57th Rifles had been sent up the high hills to the south of the valley, and were moving along abreast of the main column. The enemy did not show much keenness on this day, and most of them cleared off soon after the guns had opened fire, but the 53rd had a certain amount of fighting on the left flank, and the 21st Cavalry got home with a charge, killing twenty tribesmen. The rear-guard, 55th Rifles, had a lot of trouble before reaching camp. The Regiment was employed as baggage guard and saw none of the fighting, with the exception of one small incident. A party of five Ghazis lay up in a tumbled down hut on the hillside, about 300 yards above the road. They let the advanced troops pass them, and then opened fire on the transport. Colonel Carruthers took a party of the baggage guard with him to turn them out. The shelter in which these men were hidden was on the top of a small knoll, and Lieutenant Anderson led a party up the hill, and into the hut, only to find that it had been evacuated. The Ghazis were in a sangar a hundred yards away, engaged with the flank guard under Lieutenant Cobb (attached), who had come along the hillside. They were surrounded and three of their number had been killed. The remaining couple made a bolt for it, but ran into the arms of a flanking party under Captain Murray. They were both accounted for, and the Regiment secured a standard and five firearms as trophies.

On May 25th the Force moved south to Mulla Killa, where a good supply of running water was discovered. Here a two days' halt was made, while a convoy was pushed through from Nahakki. Hostilities had now ceased and no more troublesome nights were experienced.

The Regiment left Mulla Killa on the 28th, and finally marched into Peshawar on June 1st. Everyone had kept remarkably fit, and the Battalion marched into barracks only seven short of the number that had left camp on the 13th. This, of course, was excluding casualties.

MOHMAND EXPEDITION, 1908.

1.—Major Blakeway, General Willcocks, Colonel Birdwood, Captain Langhorne. 2.—Shabkadr Fort.
3.—Near Khan Beg Kor. 4.—Near Taraf Pass.

BRITISH AND INDIAN OFFICERS, 1907.

Colonel Cunliffe in centre.

Indian Officers seated, left to right—Sub. Bishen Singh, Sub.-Major Baz Khan, Sub. Ditt Singh.

Sub. Mohammed Khan standing on extreme left.

Considerable experience was gained from this expedition, especially in the construction of perimeter camps. The trenches dug by the 59th were always much admired, the experiences of the attacks at Matta and Garhi Sadr having shown our men the importance of good cover and also the best means of making it. The fact that in all the many night attacks nobody of the Regiment was hit in the trenches speaks for itself.

The casualties suffered by the Regiment during the expedition were :—

Killed, 1 Indian officer and 2 sepoys.

Wounded, 1 British officer, 2 Indian officers, 3 havildars, and 20 sepoys.

The following officers were mentioned in despatches :—

Lieut.-Colonel R. A. Carruthers, Captain H. de C. O'Grady, Lieutenant B. E. Anderson.

The following honours and awards were received :—

Commander of the Bath :

Lieut.-Colonel R. A. Carruthers.

Order of British India, 2nd Class (with title of " Bahadur ") :

Subadar-Major Ditt Singh.

Indian Order of Merit :

No. 4230 Sepoy Amar Singh.

At Matta on April 24th, 1908, he picked up Jemadar Jehan Dad Khan who was wounded, and although wounded himself, carried him to a dhoolie.

Distinguished Service Medal :

Subadar Makhmad Jan.

Jemadar Jehan Dad Khan.

2623 Havildar Mir Nabbi Hussain.

2635 Havildar Sansar Singh.

2894 Havildar Ralla Singh.

2917 Havildar Mobin Khan (for Zakka Khel Expedition).

3556 Sepoy Attar Singh.

3360 Lance-Naik Madat Ali.

4442 Sepoy Kaka Singh.

The following officers accompanied the Regiment on the expedition :—

Lieut.-Colonel R. A. Carruthers (Commandant).
Major E. Kirkpatrick (Second-in-Command).
Captain T. L. Leeds (up to 12/5/08).
Captain H. de C. O'Grady.
Captain H. W. Martin.
Captain K. D. Murray (Adjutant).
Lieutenant B. E. Anderson (Quartermaster).
Lieutenant W. F. Scott.
Lieutenant H. N. Lee.
Lieutenant A. H. Burn.
Captain J. Husband, I.M.S.

Attached :—

Lieutenant A. Majoribanks, 52nd Sikhs, F.F.
Lieutenant R. T. McEmery, 46th Punjabis.
Lieutenant F. F. Hodgson, 84th Punjabis.
Lieutenant C. Cobb, 29th Punjabis.

In October the Regiment suffered a grievous loss through the death of Subadar-Major Ditt Singh at his home. He was a fine straightforward Indian officer, and proved himself a fighting man of rare courage and capacity at Matta and throughout the Mohmand Expedition. He was succeeded as Subadar-Major by Subadar Bishen Singh.

The Regiment remained in Peshawar for the rest of the year. During the cold weather 1908-9 the notorious raider Multan was very active in the Peshawar Valley, and the troops were frequently called out to catch him. In January, 1909, one of his gang was bribed to betray him, and gave information to the Political Officer of the Khyber Pass, that Multan and his gang would be lying up for the day on January 23rd in a nullah near Phandu, with a view to raiding Peshawar city. The 19th Lancers were sent out to form a cordon round and capture the gang, and at 11.30 a.m. orders were received for two companies to proceed to a spot near Phandu in support of the cavalry. "D" and "C" Companies under Captain T. J. Willans and Lieutenant H. F. D. Stirling, marched at once, arriving at the rendezvous at 12.45 p.m.

Action against raiders.

No one having been left behind to show Captain Willans in which direction the cavalry had gone he did not get in touch with them until after 2 p.m., when he found that they had suffered some casualties, including a British officer, severely wounded, and had rounded up nine of the raiders in a watercut. Multan had been killed earlier in the day after a gallant resistance. Captain Willans at once advanced on the raiders, and was fired on, without suffering any loss. When his men were within a few yards of the raiders the latter asked for terms and threw down their arms.

On February 20th, 1910, the Regiment arrived in Kohat from Peshawar, on relief.

The Regiment remained in Kohat throughout 1911. There was some raiding during the autumn, and in October, information was received that a party of raiders, mostly outlaws, intended to raid Kalabagh, and would pass through a certain defile near Shadi Khel on the night of October 15th-16th. The Regiment, 500 rifles strong, was sent to Dhoda to intercept the raiders. Two hundred rifles went out to the defile to lay an ambush and of these fifty were left behind to block an alternative route. The raiders, fourteen in number, appeared at 12.30 a.m., and were challenged, but commenced running away, and Lieut.-Colonel Carruthers ordered fire to be opened. The night was very dark, and the raiders, scattering, got away with the exception of two, who died of their wounds. Another was captured at dawn by a cavalry patrol, and a fourth wounded raider afterwards came into hospital.

During 1912 the Regiment remained at Kohat. Subadar-Major Bishen Singh and Subadar Makhmad Jan, I.D.S.M., were both admitted to the Order of British India (2nd Class) with the title of Bahadur. Subadar-Major Bishen Singh retired on pension on November 1st, and was succeeded by Subadar Mohammed Khan.

On January 5th, 1913, the Regiment proceeded to the Samana, with headquarters at Hangu, and detachments at Fort Lockhart, Thal, Fort Cavagnari, Dhar, and Sangar. Headquarters moved up to Fort Lockhart on May 1st. Lieut.-Colonel P. B. Forster succeeded Colonel R. A. Carruthers, C.B., on March 2nd, and commanded until September 22nd, when he returned to the 52nd Sikhs, F.F.

Lieut.-Colonel C. C. Fenner was retransferred from the 56th Rifles, F.F., and took over from Lieut.-Colonel Forster.

On November 6th, 7th, and 8th the Regiment was relieved of the Samana posts and Thal by the 56th Rifles F.F., and concentrated at Hangu, whence the march to Jullundur was begun on November 10th. Jullundur was reached on December 17th. The total distance was 396 miles by the Grand Trunk roads. Halts were made at Kohat, Rawal Pindi, Jhelum, Lahore, and Amritsar.

In October Captains K. D. Murray, and B. E. Anderson were both successful in the competitive examination for the Staff College, Quetta.

GROUP TAKEN AT JULLUNDUR, 1914.

Standing: Lieut. J. A. M. Scobie, Lieut. W. A. McC. Bruce, Lieut. J. C. Atkinson, Lieut. R. D. Inskip.
Sitting: Captain A. H. Burn, Captain B. E. Anderson, Colonel C. C. Fenner, Captain K. D. Murray, Captain H. W. Martin.

ARRIVAL AT MARSEILLES, 1914.
Captain R. D. Inskip, Colonel C. C. Fenner, Major T. L. Leeds.

SIKH COMPANY MARCHING THROUGH MARSEILLES.

CHAPTER VI.

The Great War: 1914–1915, France.

The Regiment was in Jullundur when war with Germany was declared. The Jullundur (8th) Brigade consisted of 1st Battalion Manchester Regiment, 47th Sikhs, and the 59th Rifles F.F. The fourth regiment of the Brigade was the 28th Punjabis, who were stationed at Mian Mir.

On August 8th orders to mobilize were received, and all ranks were recalled from leave. Mobilization envelopes were sent out to all concerned. It is a fact well worthy of note that in many instances these envelopes were never delivered at all by local village postal authorities, and in some cases a fee of eight annas was demanded for their delivery. The *esprit de corps* of the Regiment was so fine that in many cases men on leave and furlough, hearing bazaar rumours as to mobilization, returned of their own accord without waiting for any official intimation. On August 13th field service clothing was issued, and reserve rations were drawn. On the 18th the Regiment entrained for Karachi; strength 9 British officers, 17 Indian officers, 1 S.A.S., 749 Indian other ranks. A depot was left at Jullundur. The following British officers proceeded with the Regiment:—

Lieut.-Colonel C. C. Fenner
Major T. L. Leeds
Captain H. W. Martin
Captain K. D. Murray
Captain B. E. Anderson
Captain R. D. Inskip
Lieutenant H. N. Urmston
Lieutenant G. C. Atkinson
Lieutenant G. Y. Thomson, I.M.S.

Captain A. H. Burn was appointed officiating Staff Captain Jullundur Brigade, and proceeded with Brigade Headquarters on August 16th.

The Regiment embarked at Karachi on the 29th on H.T. *Takada.* The 15th Sikhs, now part of the Jullundur Brigade, were on board, and also the entire Lahore Divisional Pay Staff, and part of a mule corps. The work of embarkation was carried out in fine style, and the Regiment earned the praise of the embarkation staff, who were astonished at the rapidity with which animals and stores were put on board. The ship joined the remainder of the convoy outside Karachi harbour, and the whole convoy sailed in a westerly direction, for an unknown destination on the evening of the 29th. The convoy arrived at Aden on September 6th, and at Suez on the 14th. On the following day the troops disembarked at Port Tewfik, under orders to proceed to Cairo. They remained in camp at Heliopolis until the 18th, when they left by rail for Alexandria, and once more embarked on the *Takada.* The whole fleet of over twenty transports left Alexandria harbour at about 10 a.m. on September 19th. Captains W. F. Scott and H. N. Lee and Lieutenants W. A. McC. Bruce and J. A. M. Scobie, in the meantime rejoined from home leave on September 16th, having come out from England on the H.T. *Dongola* which had been specially chartered to take back officers and civilians to the East on the outbreak of war. Each of these four officers was killed during the war.

Marseilles was reached on September 26th, and the troops disembarked the same day and proceeded to camp. All the Indian troops received a great ovation in Marseilles; the entire population turned out and cheered most enthusiastically, and strewed flowers over everyone. While at Marseilles all troops were re-armed with new high velocity rifles. The Regiment entrained at Marseilles on September 30th, and arrived at Orleans two days later, and proceeded to camp. Here the Battalion was organized in accordance with the new platoon organization, and the 1914 drill was taken into use. This had not yet been introduced into India, although troops at home had taken to it. Old double companies and half companies were abolished from this date.

The Regiment remained at Orleans until the 18th, when they entrained, and arrived at Witzernnes early on the morning of the 20th. Here several aeroplanes were flying about, and this was the first occasion on which the Regiment had a good view of them. During the night previous to entraining it rained heavily, and it was impossible to get the heavy baggage wagons out of the camp in time to catch the train, so that a large party had to be left behind, together with most of the heavy baggage wagons. These wagons were ordinary English trade wagons, drawn by fine English cart horses, accustomed to careful grooming and cleaning and good warm stables. When camped in the open, without the attendance to which they had been accustomed, they suffered greatly, and there were many casualties among them. The Indian soldiers were greatly impressed by the size and power of these magnificent horses.

During the train journey from Orleans some men of the London Scottish were met on duty at Abbeville, where they did all they could for the Regiment, producing tea, etc., in the early morning.

Arrival in the Trenches.

On October 21st the Regiment went into billets near G.H.Q. at St. Omer, and on the following day proceeded on a long march to Meteren. The men had nothing to eat all day long until rations arrived at 5.30 p.m. Outposts were put out in the pouring rain. Heavy firing was heard in the direction of Lille. At 4.50 a.m. on the following morning (23rd) orders were received for the Division to march at once to Estaires. The 57th and the Connaught Rangers had been sent on ahead in motor buses to Wytschaete, where urgent reinforcements were required by the British Cavalry Division. The march for Estaires was commenced at 5 a.m. General Sir John French was in the market square at Estaires, and the Brigade marched past him in column of route. The march was continued down the La Bassee Road to Rouge Croix, where the Regiment went into billets. The Sikh Company, under Major Leeds, was ordered to take up an outpost line beyond Rouge Croix, but the order was cancelled just as the Company was getting into position. It was here that the celebrated "coal boxes" were seen. The enemy was shelling one of our 4·7 batteries some 200 yards off the road on which the Sikh Company was situated. Many British wounded were passed, and the road was crowded with French refugees streaming back from the front. At 9 p.m. Nos. 2 and 4 Companies were sent as escort to the 18th Brigade R.F.A., and shortly afterwards orders were received to send the other two companies to relieve the Manchester Regiment at 5 a.m. Colonel Fenner was in command of these two companies, and also two companies of the 15th Sikhs, and relieved the Manchesters, and also the 47th Sikhs soon after 5 a.m. on October 24th. The Regiment dug itself in at once.

The men did not display much enthusiasm about it at first, but some enemy guns opened fire on the line shortly after the digging commenced, and a few casualties soon showed all ranks the desirability of getting under ground as soon as possible. Nos. 2 and 4 Companies rejoined at about 9 a.m. and took the place of the two companies of the 15th Sikhs, who rejoined their regiment. At dusk the Brigade Major arrived with orders for the Battalion to move forward to a line occupied by French Cavalry piquets and take over from the French. This operation was carried out after some difficulty, and a new line of trenches was dug. It is interesting to note that this line which was dug by the Regiment remained practically unchanged in British and Allied possession until lost in 1918, by a Portuguese Division.

Enemy Attacks Repulsed.

On October 25th at about 3 p.m. the enemy attempted a daylight attack on the centre of the line, held by Nos. 1 and 4 Companies. In this attack Captain Scott, of No. 1 Company, was mortally wounded. He was shot through the head while putting supports into the fire trench, and died a few hours afterwards. The attack was practically unsupported by artillery, and was easily disposed of, no enemy living to get within 200 yards of our position. The 8th British Brigade on the Battalion's right was attacked at the same time. As soon as it was dark the enemy resumed his attack, and the line was heavily pressed. It was now raining, and the new rifles began to jam. One company of the 34th Pioneers and some Sappers and Miners were pushed up as reinforcements, and by 3 a.m. the enemy had to give up and withdraw. If only he had pushed his attack home more vigorously he might have got through when the rifles were jammed. The Regiment was holding 1,500 yards of front with a strength of 600. There were no communication trenches, except such natural ditches as already existed, and platoons were dug in just where they happened to be, frequently with no lateral communication except over the top.

The following congratulatory messages were received :—

From Lord Kitchener to the Commander-in-Chief, Sir John French : " The splendid courage and endurance of our troops in the battle in which you have been engaged in the last few days, and the boldness and capacity with which they have been led, have undoubtedly given the enemy a severe blow, successfully frustrating all their efforts. Let the troops know how much we all appreciate their services, which worthily maintain the best traditions of our Army."

From the G.O.C. Lahore Division: "Have much pleasure in informing you of the following wire, which please convey to all ranks: 'Commander 2nd Corps [General Sir H. Smith-Dorrien] desires you to convey to the 8th Infantry Brigade his congratulations on the splendid fighting powers they have shown to-day. He regrets their losses, but feels sure they will maintain to the utmost their present position.'"

On the 26th the enemy were reported to be taking up strong entrenched positions opposite our centre, and closer than formerly, and early on the following day Captain Martin sent a message to say that he was being heavily attacked at close quarters, and had suffered many casualties. At the same time a heavy attack developed on the centre, and No. 3 Company, under Captain Murray, was brought up as reinforcements. Captain Murray was severely wounded in the thigh while reinforcing the centre. No. 2 Company, under Captain Martin, was reinforced by some of the 15th Sikhs, while the Khattaks moved up to reinforce the right centre. A vigorous fire fight ensued, and in an hour's time the enemy was beaten back all along the line. At this time arrangements for evacuating wounded from regimental aid posts were extremely bad. The aid posts were too far back, ours being two miles away, and stretcher-bearers had to carry men to the aid post and thence to the clearing station. There was also an order that no medical personnel were to be allowed in the firing line or any trenches. The remainder of the day was quiet except for the activities of snipers, who caused many casualties. Many were in houses and barns behind our lines, hidden well back inside the roof of the building. Later on they were discovered and accounted for.

The enemy launched two more attacks on the 28th, but were beaten off on each occasion. A night attack was repulsed on the following night, and a telegram was received from the Corps Commander conveying Sir John French's congratulations to the Indian troops on their gallant conduct. On the following day (October 30th) a wire was received to the effect that the German Emperor was arriving in the field to conduct operations against the British Army in person.

On October 31st there was more artillery fire from the enemy than usual, but no damage was done, as the trenches were good. On the following day the Regiment, with the exception of No. 2 Company, under Captain Martin, was relieved by the 47th Sikhs, after nine days in the front line. No. 2 Company had to remain behind as the 47th were not strong enough to take over the whole line. Information was received next morning that Captain Martin had been severely wounded in the wrist.

The Regiment had been in billets for barely eight hours, when orders were received to proceed at once to La Flique, and to report to the 8th British Brigade. The Germans had captured some trenches there with the aid of their Minenwerfer, heavy enfilade machine-gun fire, and bombing. The Regiment arrived at their destination at 3.15 p.m., and found that little of the true state of affairs was known. It was decided that the Royal Scots Fusiliers would counter-attack, with the 59th in support. The 2/2nd Gurkhas had been occupying a position north-west of Neuve Chapelle. This position formed a salient. Under cover of gun and mortar-fire, the Germans had been able to advance until they were within bombing distance of the Gurkha front-line trench. The Gurkhas suffered very heavy casualties, especially among British officers, and had to vacate their position. The projected counter-attack never came off, it was dark before it could be delivered. The ground to be attacked over consisted of small gardens fenced with barbed wire, and held by the enemy in great strength and with many machine guns. After long delay, and much consultation, it was decided to dig a new line behind the salient.

The trenches were begun at 1 a.m. The men had learnt the value of good trenches by this time, and by 6 a.m. all were under cover. A heavy bombardment by the enemy took place that morning, being the Regiment's first real experience of the enemy's "crump," otherwise 5.9. The shooting was accurate, and again the value of good trenches against artillery fire was conclusively demonstrated. This fire was kept up almost continuously until 4 p.m. No. 2 Company had in the meantime been relieved by a company of the 34th Pioneers.

On November 4th a fierce artillery duel took place, and on the following day the enemy attacked our position, after a heavy bombardment. The attack was beaten off, however, and that night the Regiment was relieved by the 57th Rifles and the 129th Baluchis. The Regiment proceeded to Estaires to billet in some mills, but as the caretaker of these had not been warned that any troops would be billeted there, a way had to be forced in, in the middle of the night. On November 7th the Regiment marched to billets in Picantin, as reserve to the 47th and 15th Sikhs.

The Regiment relieved the 47th Sikhs in the old original trenches on November 10th, and found that they had been improved considerably. Communications had been dug both from front to rear, and also laterally. Some hand and rifle grenades were received from the 47th, and also some sandbags, the first seen by the Regiment since arrival at the front.

The following written message was received by the Commanding Officer :—

> " I am very pleased to hear how well you are doing your duty. I know the 59th will add to its high reputation.
>
> (*Sgd.*) " JAMES WILLCOCKS,
>
> " *Lieut.-General.*"

During the next few days activities were confined chiefly to patrol work, in which Lance-Havildar Niaz Gul particularly distinguished himself. He had previously gained promotion for good scout work. The enemy appeared to be digging hard, sapping towards the 15th Sikhs, on the right. On November 10th three representatives from each unit were sent to Estaires to meet Lord Roberts. On the 15th the sad news of Lord Roberts' death was received. He died in Estaires on November 12th. Sniping continued, and hostile artillery fire was frequent. The first British officer reinforcement, Captain B. A. Trafford, 52nd Sikhs F.F., arrived on November 10th.

On the night of the 15th the Regiment was relieved by the Royal Irish, but the relief was considerably delayed and carried out with difficulty, as the enemy attacked on the right of the line while the relief was going on. The Regiment arrived in Estaires at 12.45 a.m. At 9.50 a.m. the Brigade left Estaires and arrived at La Couture some three hours later.

The Regiment remained in billets until November 21st. Time

in billets was employed in digging reserve trenches, collecting material for trenches, and in general cleaning up. H.R.H. The Prince of Wales inspected one company from each regiment at a parade held on the 31st. No. 2 Company represented the 59th.

At 11.25 p.m. on the night of November 21st the Regiment received orders to turn out at once and report to the G.O.C. Dehra Dun Brigade at Hamel. The Regiment marched at 12.17 a.m., and reported at Hamel at 1.30. It should be mentioned that the billets consisted of four or five farm buildings more than half a mile apart. First line transport accompanied the Regiment. The French on the right of the Dehra Dun Brigade expected an attack, and asked for reserves, as they had none. Dehra Dun Brigade gave theirs, which the 59th replaced. However no attack took place, and on the following morning, the Regiment returned to billets. The Divisional Commander sent a message expressing his great pleasure at the rapidity with which the Regiment had turned out.

On the afternoon of November 23rd the Regiment marched by companies to take over the Rue de Bois trenches near Port Arthur from the 2/2nd Gurkhas. The Commanding Officer and company commanders went to view the trenches by daylight. After going round the trenches Colonel Fenner went to see Battalion headquarters' dug-out; it was then 5 p.m. and dark. A stray bullet hit him in the mouth, passing through the brain, and killing him instantly. The Regiment felt his loss very keenly and all ranks were filled with deep and genuine sorrow. Colonel Fenner had spent the whole of his service in the Regiment, and knew it thoroughly, and was well liked and greatly respected by everybody. Major T. L. Leeds assumed command of the Regiment from this date. Colonel Fenner's remains were buried on the following morning. The G.O.C. Brigade and all officers who could be spared from the trenches attended. Many messages of sympathy and condolence were received from commanders of higher formations and units.

During the next few days sniping was continuous, and Lieutenant H. N. Urmston was seriously wounded in the head on November 26th. He recovered from this serious wound, but never soldiered with the Regiment again. Enemy 'planes were also active at this period, and one came over on the 24th and spotted some Dogras digging. Subsequently enemy guns opened on the Dogras, causing several casualties. The same thing happened two days later with the 15th Lancers, just behind the 59th lines, and Major Corlett was hit, as well as two of our men. Orders were given for no digging during daylight. A thaw had set in in the meantime, and the trenches fell in badly and had to be repaired. This was difficult as work could only be done at night, and it was difficult to obtain suitable materials for revetting. Enemy artillery was very active at times. Coal and warm clothing were received in the trenches. The Sub-Assistant Surgeon was ordered to treat men in the trenches with trench feet, cold, etc. This was much appreciated.

On December 2nd the Regiment was relieved by the 2/39th Garhwalis. The relief was quickly carried out and by 9.15 p.m. the Regiment was in billets at La Couture. General Sir James Willcocks inspected the Regiment on December 6th, and made many flattering remarks about the Regiment, and chatted to all British and Indian officers individually. The 4th Battalion Suffolk Regiment (Territorial) joined the Jullundur Brigade on December 4th. On December 8th a draft consisting of 2 British officers, 3 Indian officers, and 190 Indian other ranks, arrived. Many reservists, obviously unfit for service in France, were included in this draft.

On the afternoon of December 10th the Regiment took over some trenches near Givenchy from a French Territorial battalion. The relief took some time, and was accomplished with difficulty as the trench system was complicated, and the French guides did not seem sure of their way. Some of the French stayed in the 59th trenches that night, in spite of efforts to move them, as the night was very wet. After their departure thousands of rounds of ammunition were discovered in the trenches, and all their reserve ammunition was left at Battalion headquarters. The Regiment held 1,200 yards of front here, with 550 men in the firing line, and 150 in support. The right flank was on the La Bassée Canal, and the enemy trenches varied from 500 yards distance at this point, to fifty yards distance on the left flank.

On the night of December 12th a Khattak patrol was heavily fired on when near the enemy's wire, and one man was left out severely wounded. No. 3902 Lance-Naik Biaz Gul and No. 27 Sepoy Zarif Khan immediately went out to fetch him back. The wounded man, 4668 Sepoy Abdullah Khan was hit in three places, and was in great pain. His groans attracted the fire of the enemy, but he was successfully brought in though he died soon after. His rescuers were both recommended for the Victoria Cross, and though the case was strongly recommended by high authority, the much coveted decoration was not obtained in either case ; an Indian Order of Merit being awarded instead in each case.

On December 15th the Commanding Officer and Adjutant attended a conference between the G.O.C. Jullundur and Ferozepore Brigades. As the result of this conference the Ferozepore Brigade on the following day attacked the German position in front of the juncture between the Jullundur and Sirhind Brigades, due east of Givenchy. The 129th Baluchis bore the brunt of the attack of the Ferozepore Brigade, but the attack had been made without sufficient reconnaissance and preparation, with the result that there was no co-operation by neighbouring units and other arms. The 129th lost 3 British officers and 100 Indian other ranks in casualties.

Casualties of the 59th to date were :—

British officers—Killed, 2 ; wounded, 3.

Indian officers—Killed, 0 ; wounded, 3.

Indian other ranks—Killed, 26 ; wounded, 258 ; died of wounds, 5.

On the following day the Regiment was relieved by the 57th, and on December 18th the 59th were attached as reserve to the Ferozepore Brigade, being at the time in billets at Beuvry. The Commanding Officer and Adjutant proceeded to Ferozepore Brigade headquarters at Gorre for orders.

The situation was that the Corps Commander was very anxious for the Indian Corps to carry out small attacks and harass the enemy. Such attacks were almost impossible to plan in detail beforehand, and later on in the war troops detailed for such attacks or raids were withdrawn from the line and practised over similar trenches and ground dug to scale from air photographs. However, in this case, the Ferozepore Brigade was detailed to carry out an attack, and the 129th

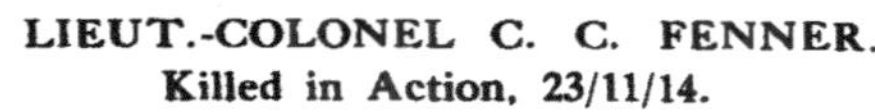

LIEUT.-COLONEL C. C. FENNER.
Killed in Action, 23/11/14.

LIEUT.-COLONEL P. C. ELIOTT-LOCKHART, D.S.O.
Killed at Neuve Chapelle, 12/3/15.

LIEUT. W. A. McC. BRUCE, V.C.
Killed at Givenchy, 19/12/14.

were ordered to carry out practically the same attack as had been attempted two days previously. The severe losses that the 129th had suffered on the 16th were then taken into consideration, and accordingly the 59th were ordered to carry out their part of the attack. It was represented that no one in the 59th had ever seen the ground before, and that as the attack had to be launched early next morning no one could ever see it as it was now dark. The order, however, had to be obeyed, and the situation and plans were discussed, rough hand sketches of the position being furnished by the 129th.

At 2 a.m. on the 19th, the Regiment marched from billets and were led by guides of the 129th into the portion of trench from which the attack was to be made. The plan of attack was as follows :—The 129th were to evacuate 200 yards of front line and support trench on the extreme left of their line, adjoining the Sirhind Brigade. The 59th were to occupy this, and in conjunction with the Highland Light Infantry and the 4th Gurkhas were to attack, capture and consolidate the German trench system in front of them. The signal for the attack was a four minutes' intense bombardment by all available guns, immediately after which troops detailed were to advance.

Nos. 1 and 4 Companies under Captain Scale and Captain Anderson were to carry out the attack, No. 1 Company on the right, and No. 4 Company on the left with a German sap running between the companies to give the direction. Great difficulty was experienced in getting out of the deep, muddy trenches.

Action at Givenchy.

No steps had been cut, or pegs driven in to help the men climb out. The sap, which was supposed to guide and divide the two attacking companies, was not discovered in time in the dark to be of any use, with the result that both companies lost direction and cohesion, but in spite of this some of them reached their objective. On the right Captain Scale reached the German wire with a platoon of Sikhs. The wire was, of course, uncut and the platoon was wiped out by machine-gun fire. Captain Scale was twice hit in the same leg, but managed somehow to crawl back to our lines. Jemadar Mangal Singh and another Sikh platoon took and held a German sap, capturing a German officer and some men. They held this for twenty-four hours until the sappers dug out to them and relieved them. On the left Captain Anderson with the P.Ms. got into the German trenches, where Germans, Gurkhas, and Highland Light Infantry were discovered all mixed up. Captain Anderson was fired at at a point blank range and hit on his revolver which was in the holster on his belt, and was knocked out for the time being, and taken back by some Highland Light Infantrymen. It was by now getting light. Nos. 2 and 3 Companies were in the support and communication trenches, and up the sap, which became jammed, making the passing of orders a difficult matter. Lieutenant Bruce with some Sikhs entirely lost direction in the dark, and eventually got into that portion of the trench held by the P.Ms. Havildar Dost Mohammed afterwards reported that he met Lieutenant Bruce there. Lieutenant Bruce was the first man into the enemy trenches, where some Germans surrendered. Lieutenant Bruce ordered Dost Mohammed to block and hold the left end of the trench. He was first wounded in the neck, and sometime afterwards was killed. While he was alive he set a wonderfully fine example to all present. Subsequently when some P.M. prisoners of war rejoined after the armistice and told the full story, Lieutenant Bruce's name was sent in for a posthumous V.C., which was granted. Havildar Dost Mohammed eventually got away when he thought that everyone was dead. As a matter of fact, many P.Ms. were then alive, but all were wounded and made prisoners.

Lieutenant Atkinson, who was with No. 4 Company, was killed while lying on the German parapet firing down into the trench with his revolver. Captain Anderson saw him there, and as soon as it was daylight his body was clearly to be seen some 250 yards away lying on the parapet, as though he were still looking into the German trench.

When it was daylight it became obvious that the attack had failed, although many men belonging to the Highland Light Infantry, 4th Gurkhas, and 59th were in the enemy's front-line trench. An effort was made to sap out to the left, but the distance was too great, and the sap could not have been run out in time. No Man's Land was swept continuously by fire of every description, and there was no chance of reinforcements reaching the enemy's trench.

Attention was now turned to the sap running from our front line to the enemy's. Captains Lee and Gilchrist had gone up this to within thirty yards of the enemy's line, and supports were sent up to hold on to what they had gained. Captain Lee was killed at this spot, and Captain Gilchrist was severely wounded. Lieutenant Scobie and Havildar Abdul Wahab did most noble work at this period, keeping the enemy back with fire, and tried to recover Captain Lee's body, but eventually had to leave it. Captain Gilchrist, who was still alive was got in with great difficulty. Captain Anderson, and Lieutenant Kisch, Royal Engineers, had by this time arrived, and the latter selected a spot to make a block, which was successfully constructed, and held by the Regiment until it was handed over to the 129th at 10 p.m. that evening. This sap was not complete, it was narrow and barely five feet deep. Great difficulty was experienced in getting wounded men out of it. In fact, the whole of the trench system was impossible to get stretchers down. The only way Captain Scale could be got out of the firing line was by lifting his stretcher above the heads of the bearers, on a level with the ground. He was exposed to fire in this position, but was got back without being hit again.

The Regiment lost 4 British officers killed and 1 wounded, and 3 Indian officers missing, including Subadar-Major Mohammed Khan.* All these were in reality killed. Indian other ranks casualties were :—Killed, 22 ; wounded, 42 ; wounded and missing, 14 ; missing, 23. The day had been a miserable one for everybody. No one had any food or blankets, and when the Battalion reached billets at Beuvry at 11.35 p.m., everyone was tired out, wet through, and in low spirits.

The Regiment was again ordered up into the line at Givenchy at 5 a.m. on the 21st, the remainder of the Brigade having preceded them the day before.

The situation at this time was that the enemy had delivered a well-planned and carefully thought out attack (as was apparent from an enemy document captured after the attack). That portion of the Sirhind Brigade line held by the Highland Light Infantry and 4th Gurkhas had been mined. There was no preliminary bombardment, and apparently nothing unusual to show that any big attack was

* Two sons of this Indian officer are serving in the Regiment at the present time (Naik Fatteh Khan and Sepoy Adalat Khan).

contemplated. The signal for the attack was the explosion of the mines, which took place at about 10.25. Practically the whole front line of the 4th Gurkhas and the Highland Light Infantry was obliterated. The 129th were driven from their trenches, and the 57th were left with their left flank unprotected. In spite of this they managed to hang on to their position. The 1st Manchesters made a most gallant attempt to retake the lost trenches, and partially succeeded in their object. It was evident that the enemy were most anxious to seize and consolidate the village of Givenchy, and they bombarded it furiously for some time, at the same time firing shrapnel at supporting trenches, inflicting many casualties. At this period the Regiment was down to 4 British officers (Major Leeds, commanding, Captain Inskip, Adjutant, and Captain Anderson and Lieutenant Scobie, each commanding two companies). The strength of Indian ranks was about 250.

News was received that the 1st Division would arrive at 1 p.m. They were much delayed. The Guards Brigade of this Division arrived at about 3 p.m., and immediately started to attack. They had received no accurate instructions, and were most grateful to the Officer Commanding 59th, who was able to explain the situation, and where things were. The Guards deployment for attack under heavy fire was carried out with the accuracy and precision of a parade movement, and was an inspiring sight. This brigade had many casualties, and the attack was never pushed home, as darkness came on before it was complete. During the night Captain Anderson did some most useful work showing various officers where different units were, and the general situation and position of troops.

On the following day Lieutenant Scobie was sent up with No. 2 Company to reinforce the 57th. Their trenches were now in a terrible state, and movement in them was most difficult owing to deep holding mud. The enemy again shelled Givenchy heavily, and the 1st Brigade pushed more troops into the front line. The I Army Corps arrived in the evening, and took over the line. The Lahore Division was withdrawn, and the Regiment proceeded to its old billets at Beuvry.

The Brigade marched via Bethune to Allouange, where they went into billets. A report of the action of December 19th was submitted, and the following names were especially brought to notice :—

Captain R. C. Gilchrist	Havildar Abdul Wahab
Captain H. N. Lee	Havildar Muzaffar Khan
Captain B. E. Anderson	Havildar Mohammed Jan
Lieutenant W. A. McC. Bruce	Havildar Dost Mohammed
Lieutenant J. A. M. Scobie	Lance-Naik Buta Singh
Subadar-Major Mohammed Khan	Sepoy Lal Khan
Jemadar Zaman Ali	Sepoy Mir Ali
Jemadar Mangal Singh.	

General Sir James Willcocks called on the Mess at Allouange on December 24th, and discussed the action of the 19th. The General afterwards talked most cordially and kindly to any Indian officer of the Regiment that could be produced on very short notice.

On December 26th the G.O.C. Division inspected the last draft that had just arrived, which contained many reservists unfit for service in France. He made a short speech in flattering terms about the work done by the Regiment all through, with special reference to the attack on the 19th. The Regiment paraded for inspection by Sir James Willcocks on the 28th. He paid special attention to the recent draft of reservists, and just as the Regiment was going to move off to billets General Sir Edmund Barrow from the India Office arrived, and he too inspected the reservists, over whom the authorities were beginning to get worried at this time.

And so ends 1914. Its long list of casualties, Colonel Fenner, Captains Scott and Lee, Lieutenants Atkinson and Bruce killed, and Captain Martin and Lieutenant Urmston wounded, and as it turned out, never to soldier with the Regiment again. On the other hand, the Regiment was making a great and glorious record, and gaining a brilliant reputation for steadiness and devotion to duty in all circumstances. Never an inch of trench lost, and ready to attack under any conditions whenever ordered.

On January 1st Captain Clerk reported his arrival for duty with the Regiment, and was posted to No. 2 Company. Captain T. Reed joined on the 5th, and on January 8th Captain P. S. Hore arrived as reinforcement. Major-General Keary arrived, and was appointed to command the Lahore Division, General Watkis having vacated some time previously. Lieut.-Colonel E. P. Strickland, after being with the Bareilly Brigade for a few days, was appointed to command the Jullundur Brigade *vice* General Carnegy, who had left on January 6th.

On January 15th the Brigade marched to La Couture, and from there the Regiment was sent to Richebourg St. Vaast, where they occupied billets and were detailed as support to the front line at Rue du Bois. While in billets they were employed in the construction of a strong post at Richebourg St. Vaast, and were subjected to some heavy shelling while at work. The district still appeared to suffer from many spies.

On the night 21st-22nd the Regiment relieved the 47th Sikhs in the front line at Rue du Bois. On arrival it was discovered that the trenches could not be occupied as they were full of water, and the position had to be held by piquets in shallow pits just behind the line. At night touch was kept between piquets by patrols. Piquets could only be relieved at night, at 6 p.m., 12 midnight, and 6 a.m., so that day piquets had to remain out for twelve hours. A drying room was provided at Richebourg, where men coming off piquet duty could get a hot bath and dry their clothes. The enemy sent up many flares at night, and had a searchlight working in this sector. Efforts were made at night to build up the second line trenches, but the mud merely slithered away, and there was no means of building up any sort of breastwork. In the afternoon Captain B. E. Anderson was hit in the knee by a sniper while walking behind the line.

The 1/39th Garhwalis relieved the Regiment on the evening of the 23rd, and companies marched independently to billets at La Couture. The Battalion remained in billets in various places until February 16th. Orders were continually received for a state of constant readiness, special activity was expected on January 27th, this being the Kaiser's birthday. On January 29th Lieut.-Colonel P. C. Eliott-Lockhart, D.S.O., arrived from India to take permanent command of the Regiment. Captain Murray rejoined from being wounded on February 2nd, and a large draft from the 52nd Sikhs, F.F., under Captain J. R. Wynter, arrived on February 9th. A fortnight after rejoining

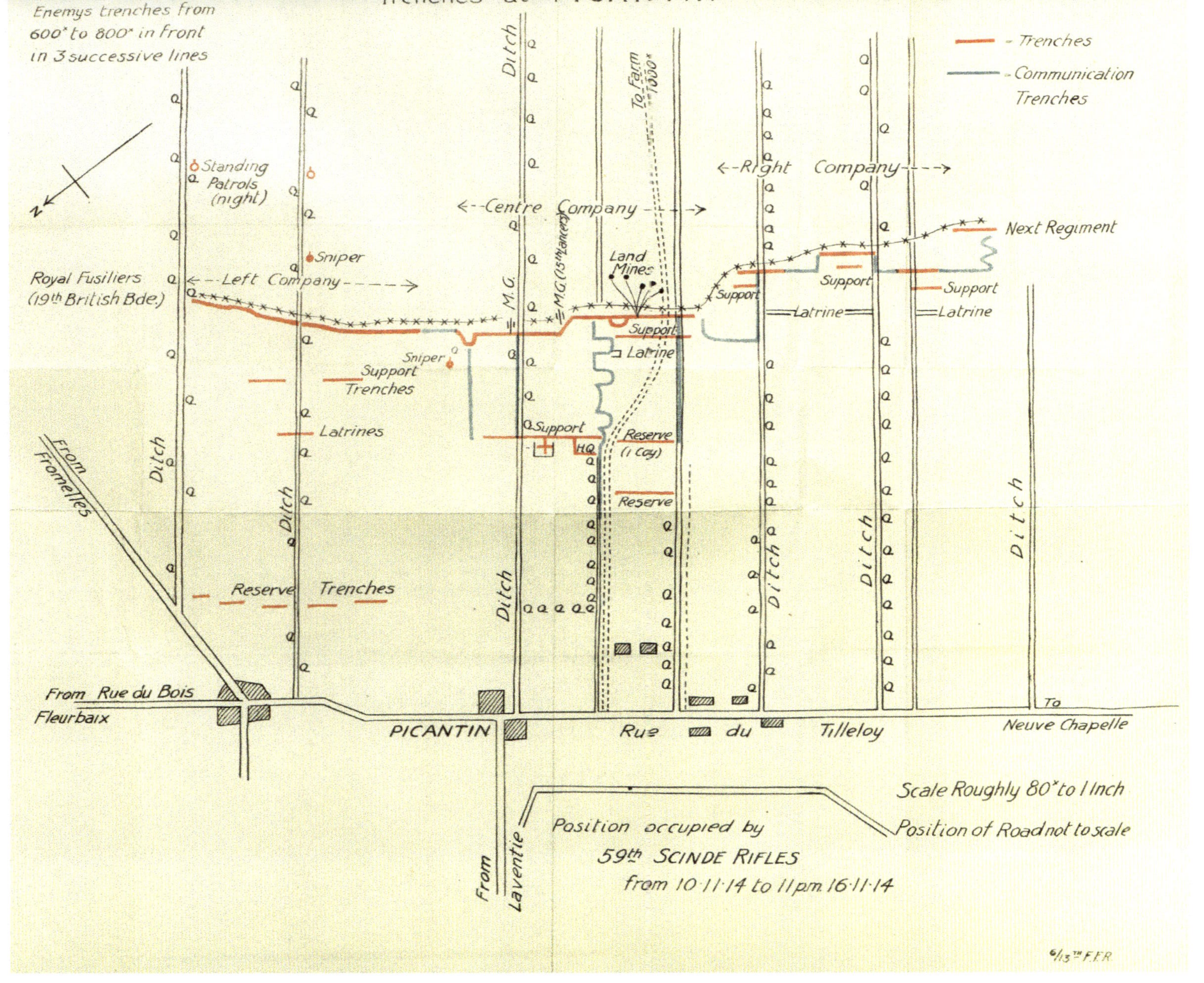
Enemys trenches from
600x to 800x in Front
in 3 successive lines
N
Standing
Patrols
(night)
Sniper
Royal Fusiliers
(19th British Bde.)
←--Left Company-----→
←--Centre Company---→
←-Right Company---→
M.G.
M.G.(15th Lancers)
Land
Mines
To Farm
1000x
Support
Latrine
Sniper
Support
Trenches
Latrines
Support
H.Q.
Reserve
(1 Coy)
Reserve
Reserve Trenches
Next Regiment
Ditch
- Trenches
- Communication
Trenches
From
Fromelles
From Rue du Bois
Fleurbaix
PICANTIN
Rue du Tilleloy
To
Neuve Chapelle
From
Laventie
Scale Roughly 80x to 1 Inch
Position of Road not to scale
Position occupied by
59th SCINDE RIFLES
from 10·11·14 to 11 pm 16·11·14
6/13th F.F.R.

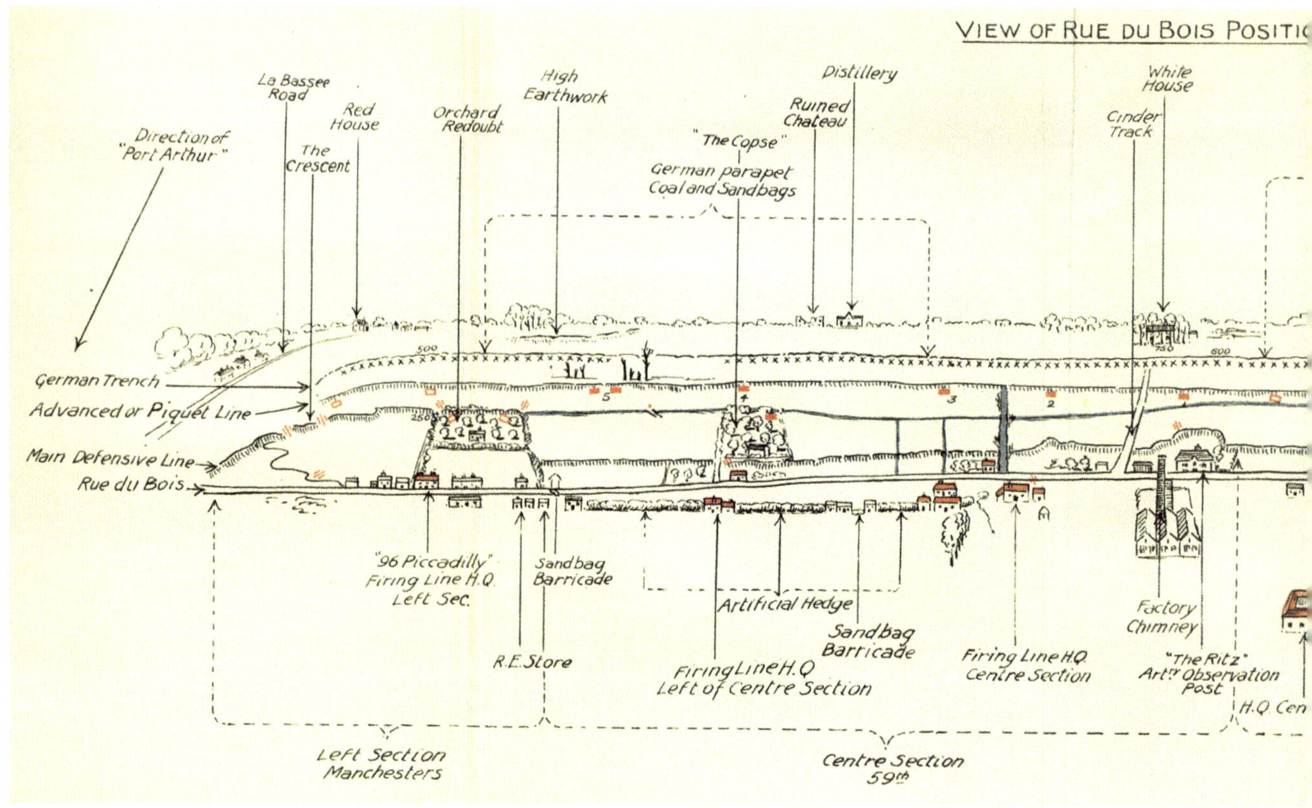
VIEW OF RUE DU BOIS POSITI
Direction of "Port Arthur"
La Bassee Road
Red House
The Crescent
Orchard Redoubt
High Earthwork
"The Copse"
German parapet Coal and Sandbags
Ruined Chateau
Distillery
Cinder Track
White House
German Trench
Advanced or Piquet Line
Main Defensive Line
Rue du Bois
500
250
750
800
5
4
3
2
"96 Piccadilly" Firing Line H.Q. Left Sec.
Sandbag Barricade
R.E. Store
Artificial Hedge
Sandbag Barricade
Firing Line H.Q Left of Centre Section
Firing Line H.Q. Centre Section
Factory Chimney
"The Ritz" Art^ry Observation Post
H.Q. Cen
Left Section Manchesters
Centre Section 59th

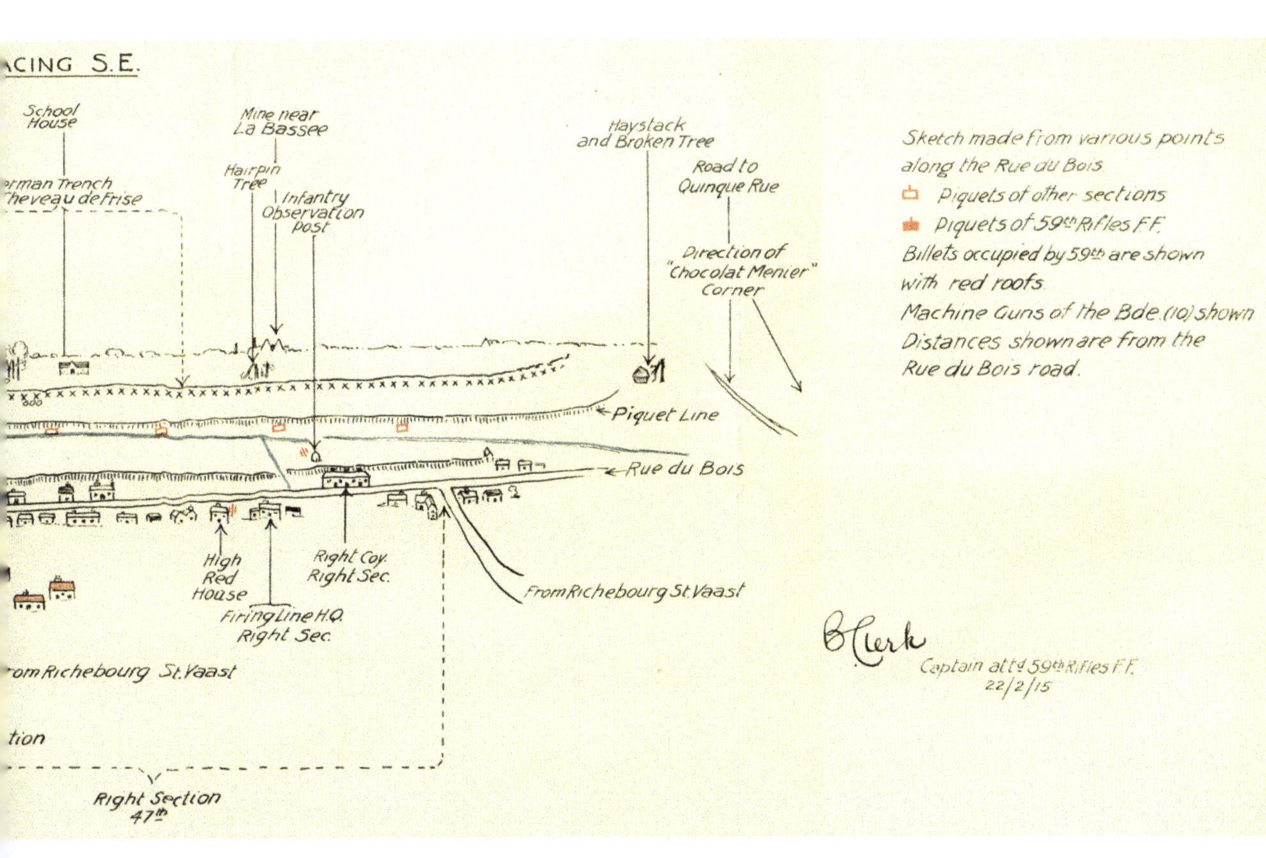
ACING S.E.
School House
Mine near La Bassee
Haystack and Broken Tree
Road to Quinque Rue
Hairpin Tree
erman Trench
Cheveau de Frise
Infantry Observation Post
Direction of "Chocolat Menier" Corner
Piquet Line
Rue du Bois
High Red House
Right Coy. Right Sec.
Firing Line H.Q. Right Sec.
From Richebourg St. Vaast
rom Richebourg St. Vaast
tion
Right Section 47th
Sketch made from various points along the Rue du Bois
Piquets of other sections
Piquets of 59th Rifles F.F.
Billets occupied by 59th are shown with red roofs
Machine Guns of the Bde. (10) shown
Distances shown are from the Rue du Bois road.
B Clerk
Captain att'd 59th Rifles F.F.
22/2/15

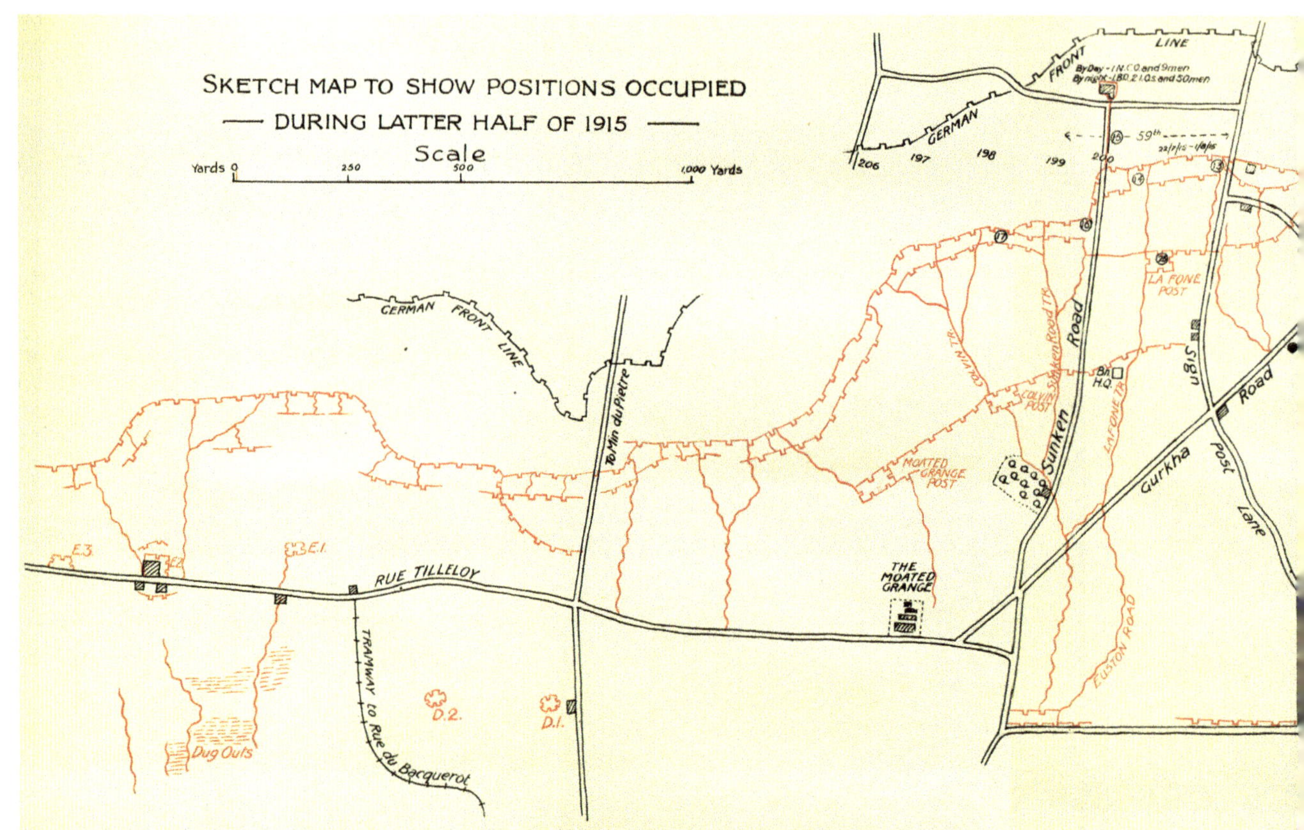

SKETCH MAP TO SHOW POSITIONS OCCUPIED
DURING LATTER HALF OF 1915
Scale
Yards 0
250
500
1,000 Yards
GERMAN FRONT LINE
By Day - 1 N.C.O. and 9 men
By night - 1 B.O. 2 I.O.s and 50 men
59th
22/7/15 - 1/8/15
206
197
198
199
200
LA FONE POST
Bn. H.Q.
COLVIN TR.
Sunken Road TR.
COLVIN POST
LAFONE TR.
Sunken Road
Sign Post Lane
Gurkha Road
MOATED GRANGE POST
THE MOATED GRANGE
To Min du Pietre
RUE TILLELOY
TRAMWAY to Rue du Bacquerot
EUSTON ROAD
E.3.
E.2.
E.1.
D.2.
D.1.
Dug Outs

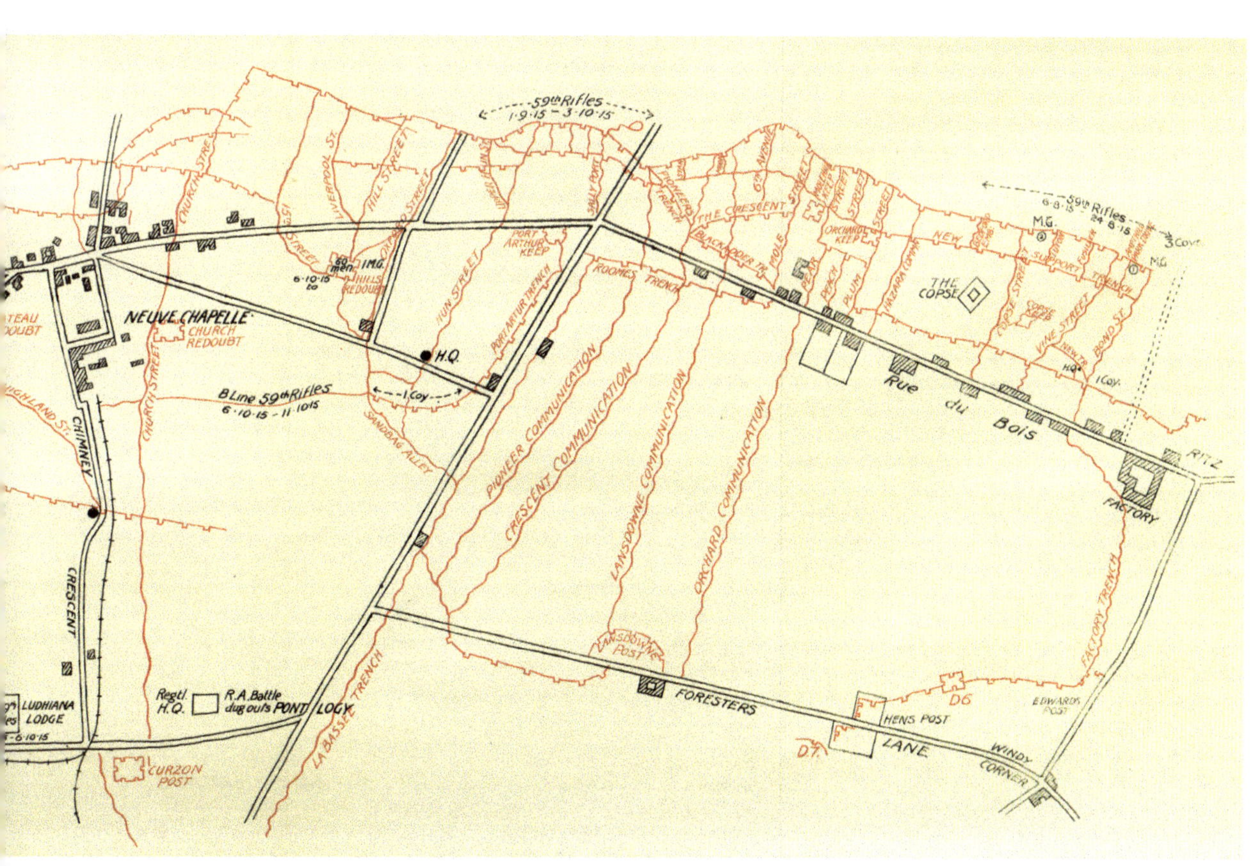

59th Rifles
1·9·15 – 3·10·15
59th Rifles
6·8·15 – 24·8·15
NEUVE CHAPELLE
CHURCH REDOUBT
H.Q.
B Line 59th Rifles
6·10·15 – 11·10·15
1 Coy
THE CRESCENT
PORT ARTHUR KEEP
ROOMES TRENCH
ORCHARD KEEP
THE COPSE
Rue du Bois
FACTORY
RITZ
PIONEER COMMUNICATION
CRESCENT COMMUNICATION
LANSDOWNE COMMUNICATION
ORCHARD COMMUNICATION
LANSDOWNE POST
FACTORY TRENCH
FORESTERS
LANE
HENS POST
WINDY CORNER
EDWARDS POST
LA BASSEE TRENCH
Regtl. H.Q.
R.A. Battle dug outs
PONT LOGY
LUDHIANA LODGE
CURZON POST
CHIMNEY
CRESCENT
CHURCH STREET
LIVERPOOL ST.
HILL STREET
HUN STREET
SANDBAG ALLEY
MOLE STREET
PLUM
NEW SUPPORT TRENCH
COPSE STREET
VINE STREET
BOND ST.
M.G.
HILLS REDOUBT

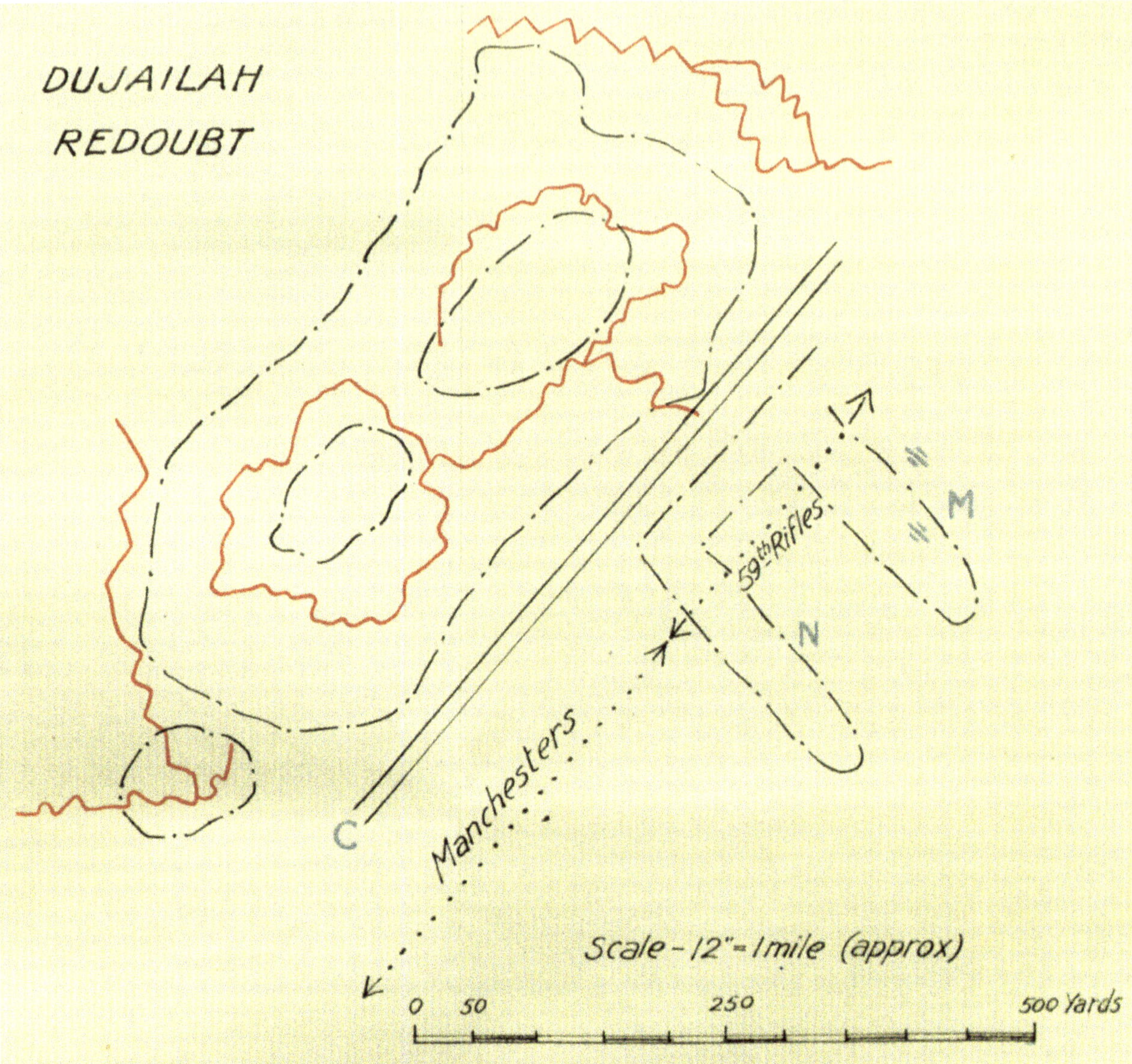

Sketch made after the action at Dujailah.

C. Canal shaped depression 30 yds. wide, possibly Dujailah depression, though its position is not as shown in T.C. 10. This depression contains water holes of good drinking water, it would also make a good place in which to collect troops for the assault, being deep and offering good cover from enemy in front

N. Nullah up which most of the 59th Rifles advanced. This appeared to be a trap as it was swept by machine gun and artillery fire, though fire was not opened until the reserve line had followed up into it.

M. Machine guns took up their position here, covering the advance of the regiment with rapid fire along the enemy's front.

Captain Murray left to take up an appointment with the Lahore Divisional Staff.

On February 16th the billets at Richebourg St. Vaast were heavily shelled, and Captain Fielding and M. Paul Ries, French interpreter, were wounded. In the evening, the Battalion relieved the 15th Sikhs in the trenches that had been held during the previous month. The Regiment remained in the trenches until relieved by the 2/2nd Gurkhas on the evening of February 23rd. On February 22nd a hostile machine gun caused much annoyance to one of our piquets. It was eventually silenced by the battery, the piquet commander observing for the gunners.

On February 28th the G.O.C. inspected the Regiment, and several awards to N.C.Os. for gallantry at Givenchy were announced.

At this time the authorities were beginning to make inquiries as to the fitness of sick and wounded men discharged from hospital. Many were sent back to their units, who were utterly unfit for duty. The G.O.C. Division inspected a party of twenty-eight such men, declared unfit by the Regimental Medical Officer, and ordered that twenty-five of these should be returned to India, and the remaining three employed on the lines of communication.

At about this time confidential notices were received to the effect that it was probable that the Indian Corps would shortly leave France. The G.O.C. Division came and discussed this with Commanding Officers and senior Indian officers of units in billets on March 9th, and impressed on all that in the meantime every effort must be made to break through the enemy line in the coming offensive. Forthcoming operations were explained at Brigade headquarters that afternoon, and in the evening operation orders were received. Subadar-Major Nasir Khan, who had succeeded Mohammed Khan as Subadar-Major, and Subadar Makhmad Jan, both long service Indian officers, whose loss was greatly felt, were evacuated to Field Ambulance and subsequently invalided back to India. Subadar Perbhat Chand became Subadar-Major.

Operations at Neuve Chapelle.

The Lahore Division was detailed as reserve in the Neuve Chapelle operations, and accordingly moved into billets, and was held in constant readiness, the Jullundur Brigade being situated at La Couture.

On March 10th, at 7.30 a.m., the bombardment by massed artillery began and continued for thirty-five minutes. Artillery fire then lengthened, and the Meerut Division attacked. The Garhwal Brigade was reported as having taken the enemy's second line trenches at 9 a.m. At 10.15 our artillery fire had slackened considerably, possibly owing to the guns moving forward. At 11.20 the Jullundur Brigade was ordered to Richebourg St. Vaast ; the roads were very congested with traffic of all sorts, but the Brigade reached their destination at 12.45 p.m. ; 183 German prisoners were passed on the road. The 1st Manchesters and the 47th Sikhs were now ordered to Neuve Chapelle. The 4th Suffolks and 59th were marched by a circuitous route to the Rue des Bercaux, and met the Manchesters and Sikhs on the road coming the other way. The road was now crammed, and the Regiment was ordered to lie down in a muddy field close to Dehra Dun Brigade headquarters. It was raining hard all this time and the road was being heavily shelled by the enemy. The Regiment was ordered to find what shelter it could for the night. During the day Neuve Chapelle and 600 prisoners were taken.

Early next morning the Brigade marched into Neuve Chapelle, and the Suffolks and the 59th were sent into support trenches thirty yards in rear of the Seaforth Highlanders. At 7.30 a.m. Neuve Chapelle was subjected to a very heavy bombardment by the enemy. The Jullundur and Dehra Dun Brigades were ordered to attack the Bois du Biez. This attack, together with that of the 8th Division, on the left, was timed for 2.15 p.m. The Dehra Dun and Jullundur Brigades moved forward, but the 24th Brigade were hung up, and their orders seem to have miscarried so that the attack could not be pressed home owing to lack of co-operation. The enemy were in the meantime searching for supports and reserves with their shell fire, and a lot of damage was done. Captain Clerk was wounded, and shortly afterwards hit again, and killed. Captain Barnes and Lieutenant Bickford were both wounded. There was now congestion in the forward area, the projected attack did not come off, and the Jullundur Brigade was ordered back to the previous night's billets.

During the night orders were received to be ready to move at 4.30 the following morning. No further orders were received, and the

1.—En route to the Front, 1914.
2.—La Couture. Winter, 1914.
3.—Richebourg Church, around which many of the 59th were wounded.
4.—Billets, 1915. Col. Eliott-Lockhart in centre.

1.—YUSAFZAIS IN TRENCHES, FRANCE.

2.—BILLET AT RICHEBOURG.

3.—MOVING UP TO NEUVE CHAPELLE, 10/3/1915.

Capt. Hore, Capt. Reed, Capt, Barnes, Lt.-Col. P. C. Eliott-Lockhart, Jem. Amir Khan and Khattacks.

Brigade moved at 5 a.m., and concentrated at Pont Logy (Rue des Bercaux) cross-roads. The enemy suddenly opened an intense bombardment on this spot, using every kind of shell, and a great many casualties occurred before the Brigade could be dispersed. The Sirhind Brigade had relieved Dehra Dun during the night, and another attack on the Bois du Biez was ordered to take place at 7 a.m. Owing to the bombardment in the dark (the accuracy and intensity of this bombardment far exceeded anything hitherto experienced) regiments and brigades had become mixed up, and it was 9 a.m. before an advance was possible. The Brigade then got going and advanced across the open under a heavy fire. During the advance Lieut.-Colonel Eliott-Lockhart was hit in the groin, and afterwards died from loss of blood. Captain Hore was killed at the head of his company, and Captain Reed was killed when bringing up the machine-gun section. Captain Burn was also badly wounded, so that the Regiment was now without a British officer, Captain Inskip having been concussed and knocked out earlier in the day by a shell. The command of the Regiment devolved upon Subadar-Major Perbhat Chand, who took over a short length of front before Port Arthur. He was later on awarded the Military Cross for his work on this day. Captain Inskip rejoined at about 8 p.m., Lieutenant Scobie at 6 p.m. from field ambulance, and Major Anderson, now recovered from his wound, at midnight. The attack previously ordered did not eventuate, and after dark the Regiment was again withdrawn to the vicinity of Lansdowne Post.

The Regiment suffered very heavily at the Battle of Neuve Chapelle, and displayed great gallantry all to no purpose. On March 11th alone there were 75 casualties, and many more on the following day, especially among British officers. During the second day Havildar Ghamai Khan displayed great coolness and resource in keeping Brigade headquarters informed of the situation on the right of the Brigade, frequently reporting in person, in spite of hostile gun and rifle fire. This non-commissioned officer rendered great service on this and other occasions during the war. He was promoted to Jemadar and afterwards received nine bullet wounds at the Battle of Dujailah from which he recovered, only to be killed while convalescing at his home (Togh), as the result of a blood-feud.

The following British officers were present at Neuve Chapelle:—

Lieut.-Col. P. C. Eliott-Lockhart, D.S.O., killed.

Capt. and Bt. Major B. E. Anderson (joined on evening of March 12th).

Capt. P. S. Hore, killed.

Capt. T. Reed, killed.

Capt. B. Clerk, killed.

Capt. A. H. Burn, wounded.

Capt. E. C. Barnes, wounded.

Capt. R. D. Inskip, wounded.

Lieut. M. H. Bickford, wounded.

Lieut. J. A. M. Scobie (joined on evening of March 12th).

On the evening of the 13th the Regiment moved from the Rue des Bercaux, and marched independently to billets at Richebourg St. Vaast. These billets were heavily shelled on the following day, and Captain Inskip was amongst those wounded. His duties as adjutant were taken over by Lieutenant Scobie.

Complimentary messages now began to pour in over the conduct of the Brigade at the Battle of Neuve Chapelle, and the 47th Sikhs and the 59th Rifles were among those mentioned in a Special India Army Order published after the battle, as having specially distinguished themselves in the fighting.

Billets were shelled again the next day, but little damage was done.

On March 18th the Regiment marched from billets and took over support trenches from the Sirhind Brigade. Some shelling was experienced, and a few casualties occurred. On March 24th the 59th were relieved, and marched to billets at Epinette, where they remained for the rest of the month. General Sir James Willcocks inspected the Regiment on the 29th, and complimented them on their behaviour in the Battle of Neuve Chapelle. Major Davis, 58th Rifles, arrived on April 1st, and took over command, and Lieutenant Campbell arrived for duty with the Regiment from the 129th Baluchis.

The Regiment relieved the 4th Gurkhas in the trenches on the 7th, and were relieved by the 125th Rifles on the 12th. During this tour, there was considerable sniping, and also some fire from Minenwerfer. Captain Turner was wounded on April 10th. The Battalion remained in billets until the 24th. Major Davis left for duty in England on the 18th, and on April 19th Lieut.-Colonel Leeds returned from sick leave in England, having been promoted Temporary Lieut.-Colonel and appointed Temporary Commandant of the 59th. On the 16th the Regiment was inspected by the G.O.C. Division, and on the following day by Field-Marshal Sir John French.

The following British officers were present with the Regiment at this time :—

Lieut.-Colonel T. L. Leeds	Lieutenant J. A. M. Scobie
Captain T. Luck	Lieutenant R. H. Burne
Captain K. D. B. Murray	Lieutenant F. A. Robertson
Captain B. E. Anderson	Lieutenant Joseph
Captain G. A. Philips	Lieutenant J. A. Shelverton

On April 23rd the Lahore Division was told to hold itself in readiness for an immediate move. That night there was a constant stream of despatch riders to and from Brigade headquarters, and at 1.30 p.m. on the following day the Brigade marched for an unknown destination. After an extremely long and trying march, Boescheppe was reached at 10 p.m. The men were all carrying over sixty pounds, and were out of condition from long spells in the trenches. Many were suffering from trench feet, and a number fell out, both British and Indian. The troops passed over the Mont des Cats before going down to Boescheppe, and obtained a view of the bombardment of Ypres. It was a wonderful sight. In the darkness, the enemy's guns, nearly surrounding Ypres, were bombarding that place furiously, and the explosions of the shells made nearly a complete ring of fire round the city. Our artillery was replying, but put up a small bombardment compared with that of the enemy. The Brigade marched again at 7 a.m. the following morning and reached Ouderdom at 10 a.m., having accomplished over thirty miles in twenty-one hours. On that evening orders were received to be prepared to march at 6.30 a.m., and to go into action at the end of the march.

Fighting in the Ypres Salient.

The Brigade marched at 6.45 a.m. on April 26th and followed the road leading round the southern edge of Ypres. Commanding Officers were called up to see the Brigadier, who had up till then received no operation orders, and could only inform Commanding Officers that they were to concentrate at Wieltje. A distance of 500 yards between units was ordered. As soon as the march was resumed the enemy shelled the road with 11-inch howitzers, 5.9s, and every kind of high-explosive and shrapnel. The 40th Pathans marching immediately in front of the 59th, received one 11-inch shell in the first line transport, which accounted for 25 men and several mules. The Regiment reached Wieltje with very few casualties. The whole road was strewn with dead horses and the Brigade was now under heavy shell fire from north, east and south.

At 12 noon the Brigadier, with Commanding Officers round him, was still awaiting orders. These were received some twenty minutes later, and the Jullundur Brigade was ordered to attack due north, together with the Ferozepore Brigade, at 1 p.m. The position for deployment was nearly a mile away, and orders had to be hastily issued. The Brigade moved off almost immediately to the place of deployment. The order of battle was—Manchesters on the right, 40th Pathans in the centre, and 47th Sikhs on the left, in two lines 200 yards apart, with men extended to three paces. The 59th were in rear of the Manchesters, and the Suffolks were in rear of the 47th, at a distance of 300 yards, also in two lines. Direction of the attack was due north. The Regiment deployed in a most steady manner, the movement being carried out as if it were being done on the parade ground. Lieutenant Scobie, Acting Adjutant, was hit by shell fire during deployment. The advance for the first 1,000 yards was up a grass incline to the top of a ridge. No enemy could be seen, but it was obvious that our troops were in full observation, as they were enfiladed by gun, machine-gun and rifle fire all the way up the slope. The attack went at a great pace, and as soon as the top was reached, a line held by all kinds of mixed units was found, and the enemy could be seen in position on a ridge opposite at a distance of about 1,000 yards, there being a valley in between the two positions, with a stream at the bottom. The attack swept over the valley, suffering severely, and part of the front line gained cover behind farm buildings and in an old trench within 200 yards of the enemy position. Parts of the front line were broken by the terrific cross fire of machine guns, rifle and artillery. The French and Ferozepore attacks on the left were also held up. The enemy was using gas shells, the Regiment's first experience of such. The 5th Division attacked later, but with no better success.

It was then realized that this daylight, unreconnoitred attack had now failed, and the G.O.C. Jullundur Brigade held up the support line and reorganized on top of the ridge. The falling back of the front line caused some confusion in the trench, but this was soon rectified; crowds in the trench were dispersed, improvements to the defences were started, and preparations for counter-attack taken in hand. From 3 p.m. the trenches the 59th were holding were being enfiladed methodically, but surprisingly few casualties occurred. Captain Murray was at this stage hit through the top of the scalp. Lieutenants Joseph and Shelverton were also wounded.

After darkness large numbers of wounded and gassed men were continually crawling back to this trench. The regimental stretcher-bearers worked unceasingly throughout the afternoon and night, carrying men of all units back to the aid post, nearly a mile in rear. During the afternoon the Highland Light Infantry and 4th Gurkhas were brought up in rear as supports.

At dawn on April 27th the Jullundur Brigade was relieved by the Sirhind Brigade, and the Brigade bivouacked at La Brique, where they remained in readiness to reinforce an attack which was to be made at 1.15 p.m. Shelling continued, and some casualties occurred. Ypres was bombarded with 17-inch shells, which passed directly over the bivouac, situated 1,000 yards in front of Ypres, for several hours during the day. One shell was fired punctually every five minutes. Heavy shelling continued all the next day. At 8.30 p.m. the Brigade marched for Ouderdom by the road north of Ypres, and were heavily shelled by what was said to be a naval gun, mounted on an armoured train. At first the shells were fused too long, but they were shortened, and one shell caught No. 4 Company, hitting 18 men, of whom three were killed. There was more shelling farther along the road, but no more casualties occurred, and the Brigade reached Ouderdom at 10.20 p.m. They had to move on from here to a position some one and a half miles away, owing to heavy shelling, by which the transport had been stampeded, and on the following day (May 1st) the Brigade moved to

Fletre and then to Croix Marmeuse, where billets were entered at 2.30 a.m. on May 2nd.

This was the end of the 2nd Battle of Ypres.

On May 4th the Regiment was again ordered up into the trenches, and relieved the Garhwal Rifles in the Neuve Chapelle trenches. The relief was accomplished with difficulty, and was delayed, as the Garhwali guides were not forthcoming, the night was pitch dark, and the roads were crammed with men and transport. The usual sniping and shelling took place, and on May 8th news was received that the Meerut Division would be attacking on the right on the following day, and that co-operation by fire on the Bois du Biez was required.

A heavy bombardment by about 500 British guns was carried on from 5 a.m. for forty minutes, when the infantry assault began. The 4th Corps gained a line of trenches near Festubert, but the attacks of the 1st Division and Indian Corps were repulsed. Neuve Chapelle and neighbourhood was heavily bombarded by the enemy. At 4 p.m. the Meerut Division again attacked, but without success. Orders were given for a night attack, but were afterwards cancelled. The Ferozepore Brigade was brought up in close support of the Jullundur Brigade. The enemy kept up heavy shelling for the next few days.

On May 15th the Regiment was ordered to co-operate by fire in a night attack. The attack began at 11.30 p.m., preceded by a bombardment. The 2nd and 7th Divisions gained some ground, but the Meerut Division had no luck. Fire was kept up by the Jullundur Brigade on the Bois du Biez until midnight. At 3 a.m. there was a furious bombardment from both sides. At dawn the enemy devoted most of his attention to the 2nd and Meerut Divisional fronts. Neuve Chapelle was shelled by the enemy off and on the whole day. Shelling continued, and on the evening of May 18th the 59th were relieved by the 129th Baluchis, and moved into billets near Estaires.

The events of the last four weeks had strained everyone's nerves to breaking-point, and all stood in need of rest. During this rest gas-respirator drill was speeded up and practised assiduously by all ranks.

On May 29th the Regiment relieved the 4th King's Liverpool Regiment in trenches at Edgware Road. This tour was fairly quiet.

General Carnegy, commanding the Jullundur Brigade in France, with Major Hill, Brigade Major (afterwards killed).

Jullundur Brigade refilling point, La Calonne, 1915.

MUG'S HOLE, NEUVE CHAPELLE.
Occupied by Battalion Headquarters, May, 1915.

The Regiment put in some very good patrol work, until relieved by the 89th Punjabis on June 13th. Captain R. D. Inskip returned, after being wounded, on June 6th, and resumed his duties as Adjutant. The Battalion was inspected in billets on the 14th by the Corps Commander, who expressed his pleasure at seeing the Regiment once more and complimented all ranks on the good work done. Information was received that Captain Murray had been given Staff employment with Kitchener's Army, and that Major Anderson had been appointed Brigade Major, Dehra Dun Brigade. Captain Phillips reported his departure for duty with the 129th Baluchis. On June 20th " smoke helmets " were issued.

The Regiment relieved the 1/4th Gurkhas at Windy Corner on the evening of the 20th, as brigade reserve some 900 yards behind the firing line. The strength, in spite of large drafts, was now down to 6 British officers, 10 Indian officers, 185 Indian other ranks, and 16 in the Machine Gun Section. They were relieved here by the 2nd Leicesters on the 24th, and on the following day took over from the Royal Berkshire Regiment in the Rue de Baquerot trenches in the neighbourhood of Chapigny. Lieut.-Colonel Leeds was in command of the left sub-section of the line, composed of the 4th Suffolks and the 59th Rifles.

The 59th were relieved by the Manchesters on June 29th, and returned to billets at Croix Marmeuse. During this period in the trenches nothing special happened. The usual shelling was experienced, and an enemy mine was detected and blown in in front of the Suffolks. A gas attack was rumoured on the night of the 27th, but nothing happened. The Regiment remained in billets at Croix Marmeuse until July 14th, and then moved to billets at Regnier Le Clerc, where large digging fatigues behind the line were supplied. Monsieur Paul Ries, French Interpreter, proceeded to St. Omer, where he was invested with the D.C.M. by Prince Arthur of Connaught, a well-merited reward. Lord Kitchener inspected the Jullundur Brigade on July 7th. Information was received that Subadar Makhmad Jan had been awarded the 1st Class Order of British India, with title of " Sirdar Bahadur," and that Captain Murray had been appointed to the Staff of the 2nd Canadian Division at Shorncliffe. On July 10th a party of fifty Indian ranks marched under Captain Inskip to Estaires, where the G.O.C. Division presented medal ribbons to all men of the Division who had been granted decorations.

On July 16th two British officers, 4 Indian officers, and 245 Indian other ranks arrived. A good draft, containing 52nd, 59th, and 25th Punjabi men.

The Regiment returned to Croix Marmeuse on July 21st, and moved up to relieve the 4th Black Watch in the trenches on July 23rd. These trenches were situated in front of Lafone Post, and included the celebrated Duck's Bill. The Regiment remained here until relieved by the 57th on July 31st. Some good patrol work was done, especially by Subadar Bishen Singh. Both sides remained fairly quiet, improving trenches, and there was a good deal of mining going on with a view to the future. The Regiment went into the trenches again on August 8th, on the Rue du Bois, where they remained until August 24th, and then moved into billets. The usual trench routine was carried on. The enemy were suspected of mining, and the Brigade bomb gun fired at the suspected mine-head. This drew heavy fire from the enemy's Minenwerfer and rifle grenades.

After seven days' rest the Regiment again went into the trenches on September 1st, and remained there for the whole month, a most trying period. Lieut.-Colonel T. L. Leeds was in command of the sub-section held by the 59th and the 47th, the 59th holding the line La Bassée Road inclusive on the right, and Oxford Street, exclusive on the left. The enemy showed activity in an orchard opposite the sub-section, and this was frequently attended to by our guns to stop their working parties there. Captain Luck shot a German who was outside the parapet near this orchard on September 6th. Several rifles with telescopic sights and periscopes had been issued to the Regiment, and sentries were always on the look-out for a chance to shoot an enemy by day. Some parachutes were experimented with for carrying lights over the enemy's lines at night. One of these was very successful. On the 10th a German aeroplane was brought down by a direct hit from one of our anti-aircraft guns. The usual trench routine was carried out, and the enemy occasionally was very active with his Minenwerfer. As before, good patrol work was constantly done by the Regiment. Internal reliefs were carried out between companies.

On September 11th an officer's patrol under Captain Fielding and Lieutenant Scobie was sent out to make a reconnaissance of the ditches in front to find out what bridging material would be necessary to enable artillery to get across in the event of an advance. This reconnaissance was successfully carried out and the patrol was returning to the front line, and had not a dozen yards to go, when Captain Fielding was shot through the heart and killed instantly. This officer had been with the Regiment since January and was a very great loss. He was universally popular, and a most keen and thoroughly efficient soldier in every respect.

On September 12th General C. A. Anderson, the new Corps Commander, came round to this part of the line. He was well known to the Regiment as he had commanded the 1st Infantry Brigade in Peshawar when the Regiment was there, and the 59th had served with him in the Zakka Khel and Mohmand Expeditions.

On September 20th all Commanding Officers were summoned to a conference at Brigade headquarters. There was an idea that a big attack would take place soon. The French guns to the south had been bombarding continuously for two days, and our own guns started on the 21st. The enemy were replying, but had done no damage so far. On September 24th a draft of 2 Indian officers and 86 Indian other ranks under Lieutenants Young and Moody, both I.A.R.O., arrived.

Orders for operations on the 25th were issued. The Meerut Division was to assault on the left of the Jullundur Brigade, who were to co-operate with fire and exploit any chance of success as opportunity occurred. The Brigade also had to make a smoke screen, but this was not a success as the wind was unfavourable. The enemy retaliated with a heavy bombardment and the front line was hammered by field artillery. Battalion headquarters was attended to by 5.9 and 4.2, and all approaches were kept under a barrage.

Battle of Loos.

This was the big Loos offensive. The action of the Indian Corps was merely a blind to keep the enemy pinned to his ground and prevent him reinforcing farther south. The Meerut attack, the signal for which was the explosion of a big mine near the Duck's Bill, was at first a great success, the Black Watch and the 58th penetrating a long way to the Aubers Ridge. Had this attack had depth and cohesion it would have been a complete success, and the Aubers Ridge and the Bois du Biez would have been captured. Heavy rain set in at 4 p.m. Orders were issued for patrolling between 2.30 a.m. and 4 a.m., to keep the enemy awake, and prevent them from repairing their parapets. Our artillery fired intermittently throughout the night. At dawn on the 26th there was a heavy mist, and not a shot was fired.

The same procedure was adopted on the night of the 26th, and there was constant gun and rifle fire. This was the anniversary of the Regiment's landing in France. There was only one Indian officer and about fifty other ranks who had been present every single day with the Regiment since landing. On October 1st the enemy put up a tremendous bombardment on Port Arthur, held by a company of 59th, with 5.9 shells. He put seventy-five of these into this post, but the garrison evacuated it while the shelling was going on, and there were no casualties. Captain Stirling was wounded while visiting the Duck's Bill on October 9th.

On October 11th the Jullundur Brigade was relieved by the Ferozepore Brigade, and the Regiment proceeded to billets after having been forty-one days in the trenches, the longest spell anyone had had up to that time. On October 14th Lieutenant J. S. Culverwell reported his arrival. The Regiment was inspected by the G.O.C. Division on the 23rd.

On October 27th the 59th took over front-line trenches from the 87th Punjabis, just south of Neuve Chapelle, and remained here until November 4th, when they were relieved by the 1/1st Gurkhas.

On October 31st intimation was received that the Indian Corps would probably embark shortly at Marseilles, the Lahore Division beginning to entrain on about November 9th. The relief by the 1/1st Gurkhas on November 4th was considerably delayed owing to a thick fog, and also the terrible condition of the trenches after constant rain. The Regiment marched to billets at La Gorgue.

This finishes the story of the 59th Rifles F.F. in the trenches in France.

The Regiment had added vastly to its reputation since the first shell burst near it on October 23rd, 1914, at Estaires. The Regiment had not lost a single yard of trench which it had been told to hold, and every time the Regiment had been ordered to attack, it had done so with the utmost gallantry.

On November 16th Brigadier-General E. P. Strickland, C.M.G., D.S.O., was transferred to command the 89th Brigade. He had commanded the Jullundur Brigade with great success since January, 1915. The following is an extract from a letter he wrote to Lieut.-Colonel T. L. Leeds :—

"I can't tell you what I feel about you and your Battalion. Throughout all this turmoil I have felt that they were an asset and to be trusted, and you can realize what that meant to me sometimes. I am more grateful to you than I can say for the assistance and support you gave me. The very best of luck to you and your splendid lot. There is no doubt about their doings if they ever get a chance, which they will."

On November 25th detachments from all units of the Indian Corps paraded at Corps headquarters for inspection by H.R.H. The Prince of Wales, who read a message of thanks for the Corps' services in France from His Majesty The King.

The Regiment entrained for Marseilles at Berquette on December 10th, and arrived at Marseilles on the 13th, when they embarked on H.T. *Canada.* The 89th Punjabis and 3rd Divisional Signal Company also embarked. Captain L. M. Heath with 1 Indian officer and 82 Indian other ranks also embarked. He had brought a Yusafzai Company of 250 men to join the Regiment, but the remainder of these, much to their disappointment had to be left behind, as the Battalion was not allowed to embark over strength. Information was received that Subadar-Major Perbhat Chand had had the honour of being invested with the Military Cross by His Majesty The King at Buckingham Palace. The H.T. *Canada* sailed from Marseilles on December 14th, 1915.

The following British officers were killed in action with the 59th in France during 1914-15 :—

Casualties in France, 1914-15.

Lieut.-Colonel C. C. Fenner.
Lieut.-Colonel P. C. Eliott-Lockhart.
Captain R. C. Gilchrist.
Captain W. F. Scott.
Captain P. C. Hore.
Captain H. N. Lee.
Captain H. C. Fielding.
Captain T. Reed.
Captain B. Clerk.
Lieutenant J. C. Atkinson.
Lieutenant W. A. McC. Bruce (afterwards awarded V.C.).
Died—Lieutenant E. Wainwright.

The following British officers were wounded :—

Brevet Major B. E. Anderson.
Captain K. D. Murray (twice).
Captain J. D. Scale.
Captain H. F. D. Stirling.
Captain E. C. Barnes.
Captain H. W. Martin.
Captain A. H. Burn.
Captain H. C. Fielding.
Captain R. D. Inskip.
Captain H. G. Turner.
Captain H. N. Urmston.
Captain M. H. Bickford.
Lieutenant J. A. M. Scobie.
2/Lieutenant A. F. Joseph.
2/Lieutenant J. A. Shelverton.

Indian Officers :

Killed—Jemadar Hukmat Khan.
Died—Jemadar Niaz Gul.
Wounded and missing—Subadar-Major Mohammed Khan.
Jemadar Zaman Ali.

Missing—Jemadar Maggar Singh.

Wounded—Subadar Vir Singh.
Subadar Chattar Singh.
Subadar Sahib Singh.
Subadar Khan Gul.
Subadar Bachittar Singh.
Subadar Mangal Singh.
Subadar Zarin Khan.
Jemadar Amin Khan.
Jemadar Gulab Din.
Jemadar Natha Singh.
Jemadar Chunak Singh.
Jemadar Ali Mohammed.
Jemadar Santa Singh.
Jemadar Gulab Singh.
Jemadar Jehan Dad Khan.

Rank and File :

Killed—71.
Died of wounds—31.
Died of disease—12.
Missing—53.
Wounded—599.

Followers :

Wounded—2.

Total casualties, all ranks, while serving in France—815.

CHAPTER VII.

THE GREAT WAR: 1916–1917, MESOPOTAMIA.

DURING the voyage through the Mediterranean there were elaborate precautions against submarines, which were very active at that time. The transport called at Alexandria on December 19th, and at Port Said on the following day, where secret orders were received for the ship to proceed to Basra. Precautions against sniping while going through the Suez Canal were taken, and Suez was reached on the 22nd. H.T. *Canada* was too big to get over Basra Bar, so that men and stores had to be transhipped on arrival there onto the *Nizam*, which arrived at Basra on January 4th, 1916.

First experiences in Mesopotamia.

The Regiment disembarked the next day, and proceeded to Makinah. There was nobody there to meet the troops, and no arrangements had been made. On the 7th orders were given for the Regiment to march, but these were later cancelled as up to the present stores had not been got off the *Nizam*. The Regiment had no transport, and there seemed to be no arrangements for disembarking stores. Much equipment remained to be drawn from arsenal. Lieutenant Stuart Prince joined with a draft from the Depot. At 3.30 p.m. on the 8th, the Regiment marched as part of an echelon, the other units being the 89th Punjabis and a medical unit. Lieut.-Colonel Leeds was commanding the echelon. The march up the Tigris proved to be most trying. Very meagre information was given as to the route to be followed, faulty maps were supplied, and the heavy kits of the echelon were carried up the Tigris in *mahelas*, which had a habit of disappearing for days at a time. This meant that the echelon frequently had to bivouac and as it was the rainy season, much discomfort ensued. There was no sign of any track, and owing to the heavy rains, marching was most difficult, frequently two miles an hour being all that could be managed. The troops were not in the best condition, after two weeks on board ship, and falling out was common at the beginning of the march. All the *mahelas* were collected by January 13th. On the 19th Amara was reached, the going being as bad as ever. On leaving Amara, the echelon was held up after having marched about ten miles, by floods. The river had burst its banks, the echelon had to camp on the spot, and construct bunds on the river bank and western face of the camp. Rain continued all that day, and everyone was soaked to the skin. Arabs were working on the gap in the river bank, and succeeded in closing it on the 23rd, with the assistance of large fatigues from the echelon. A path over the bund, wide enough to take men in single file was ready, but the mules and transport had to make a big detour through swamps, and were unfit to proceed farther when they joined the echelon, which in the meantime had advanced only three miles. On the 25th the echelon caught up 3rd Echelon, who were held up at Koweit, the river having again burst its banks. General Sir Percy Lake arrived on the same day on his way up to the front. He was horrified at the meagre information and instructions that had been given to the Officers Commanding echelons, and at the inadequate maps supplied. Further advance was possible on the 27th, but transport rations were running low, owing to the delays, and a *mahela* had to be sent back to Amara to get up fresh supplies. Further delays were caused by a head wind, which obstructed the progress of the *mahelas*. These had to be towed along by fatigues, and even then could only make about three to four miles an hour. Short marches only could be accomplished, and even then *mahelas* got in late at night. The transport also had a bad time getting through the swampy ground. These conditions continued until February 3rd, when Minthar was reached, and some semblance of a track was discovered. This made the going much easier, and eventually the Jullundur Brigade reached beyond Orah, on February 6th. The Jullundur Brigade now consisted of the 1st Manchesters, 4th Rajputs, 47th Sikhs, and 59th Rifles. On February 3rd Captain T. Luck reported his departure to command a wing of his own regiment, the 67th Punjabis, at Ali Garbhi.

On February 12th the 8th Brigade relieved the 9th Brigade in the trenches. The 59th took the place of the 1/1st Gurkhas in an outpost line, running in a semi-circle, with the right flank, held by the

47th Sikhs, resting on the Tigris. The position was liable to enfilade from the enemy on the other side of the river. Men were employed in improving the trenches. On the 15th Captain J. B. Gordon, 52nd Sikhs F.F., arrived. On the following day all men of the 56th returned to their own regiment, and the 59th lost four good non-commissioned officers and forty-four men, many of them specialists.

On February 20th a forward movement was discussed, and on the following day Lieutenant Scobie took out a small patrol to reconnoitre the ground over which the Regiment was to advance. The Brigade fell in in front of the trenches that evening, and marched in a southerly direction to join the 7th and 9th Brigades at Sanna. At 2 a.m. on the 22nd, the 3rd Lahore Division and three batteries of artillery moved off west, and then swung north, the object being to open sudden rapid fire on the Turkish camp across the river. The 59th were detailed as escort to the guns while on the march and in action. The march was only about three miles, but marching in the dark was slow and tedious. At 5.45 the guns turned north, and the 59th took up a covering position about 500 yards in advance of the guns and dug in there. Arab piquets and patrols were encountered and fired on, and Captain Inskip managed to capture an Arab who was running away, and made him a prisoner, tying him up with his pugri. The 1/1st Gurkhas had been detailed to capture some mounds afterwards called Mason's Mounds, but the 59th had included this in part of their covering position for the guns. The party there was fired on by the Gurkhas in the dark, but the mistake was recognized at once and no harm was done. The guns opened a heavy fire at 6 a.m., and seemed to be doing great execution in the enemy camp across the river. The enemy guns at once began to reply, and the 59th received their full share, two men being killed, and six wounded. The transport, which was too close up, also suffered, and many mules were hit. At 1 p.m. orders were received for the Regiment to rejoin the Brigade one and a half miles to the south. On arrival there they dug in as Brigade reserve. At 5.15, when the digging was completed, and men were settling in for the night, a fresh order was received to take up a piquet line between the 8th and 9th Brigades. This meant moving another one and a half miles. The Regiment was bombed by a hostile 'plane on moving off, but no damage was done.

At 6.30 p.m. the new piquet line was taken over, and the Regiment dug in on a frontage of 1,000 yards, each company having three piquets of one platoon each, and one platoon in support. The last twenty-four hours had proved very trying and exhausting. No one had had any sleep, and the men had dug in three times.

On February 24th patrols were sent out but could see no signs of the enemy to the immediate front. The 8th Brigade (less 59th) moved out in support of an artillery brigade, which shelled the Sanna-i-Yat position. In the afternoon the Regiment was ordered to close and await orders from the 7th Brigade. No written orders arrived, but eventually a vague verbal message was received from the 8th Brigade to march on a certain compass bearing from a spot unknown, and help the 7th Brigade dig a new position. The order was complied with as well as possible, but the Regiment took over three hours in discovering the 7th Brigade. A guide would have saved two and a half hours' marching, carrying greatcoat, blanket and entrenching tools on the man. Here the Battalion had to dig 160 yards of trench, which was quickly done, and once more returned to the piquet line. On the following day the Regiment was ordered back into Brigade reserve. The artillery again shelled the Sanna-i-Yat position, supported by the 8th Brigade (less 2nd Rajputs). All this time enemy 'planes were very active.

The trenches in the piquet line were filled in on the 27th, and the Brigade moved out about three miles to support a cavalry brigade, who were to make a reconnaissance of the Sinn position, and then try to draw enemy cavalry across the 8th Brigade front. Enemy cavalry followed them up on retirement, but did not come near enough for any fire to be opened on them.

On the 28th the 8th Brigade relieved the 9th Brigade. The 59th trenches on Mason's Mounds were badly enfiladed, and casualties occurred. Trenches had to be improved and traverses made more substantial.

Attack on Dujailah Redoubt.

Preparations were now made for the operations of March 8th. Lieutenant Scobie went to Divisional headquarters on the 5th to reconnoitre the road to the rendezvous. A fatigue party made a road there on the following day. A conference was held at Brigade headquarters on the evening of the 6th. The Divisional rendezvous was reached at 8.30 p.m. on the 7th. The force, consisting of six infantry brigades (it was not treated as two divisions, command was centralized), seven batteries of artillery, and one cavalry brigade, eventually moved off at 10.15 p.m. The general idea of the operation was to attack the Es Sinn position from the west, sending two infantry brigades and one cavalry brigade round the enemy's right flank. The force marched all night, and attacked at dawn. The 8th Brigade were detailed as reserve, and the 59th as escort to the guns. They took up an extended position, and dug themselves in on the right of the guns at 7 a.m. The Regiment was ordered to concentrate at Brigade headquarters, and after having to pass through heavy shell fire by platoons, arrived at Brigade headquarters at 11 a.m. At 12.45 the 8th Brigade was ordered up to the support of the 7th Brigade, one mile north of the Sinn Abtar Redoubt. This movement was completed by 1.20 p.m., and regiments had again dug themselves in. Shortly after the 8th Brigade was ordered to move round behind the guns and support the 37th Brigade. By 3 p.m. the Brigade had moved round and formed up under cover of some mounds about 3,500 yards east of the Dujailah Redoubt. The Regiment had now been on the move for twenty-four hours and had twice dug in.

At 4.15 p.m. the 8th Brigade was ordered to attack the Dujailah Redoubt. The Officer Commanding 59th was informed that three other brigades were taking part in this attack. The Brigade was disposed for attack as follows :—Firing line and supports, 59th on the right and Manchesters on the left, two companies each in front, and two in support. The 2nd Rajputs followed in support of the above two battalions, and the 47th Sikhs were in reserve.

There was a preliminary bombardment by artillery, but as the Dujailah Redoubt was situated due west, and the sun was getting low, observation was impossible. As soon as the bombardment was over, the attack started. It went at a great pace, five miles an hour throughout, with no halts. Distance and direction were excellently maintained. After advancing about 800 yards the leading battalions came under very heavy gun and rifle fire, and casualties were numerous, but the pace never slackened. The 59th line of advance lay up a low depression with banks on either side, which looked as if it would afford cover, but it did not, and was in fact a trap. Many casualties occurred here, and Lieut.-Colonel T. L. Leeds, who was with the support line, was wounded here. The 59th lost touch to some extent with the Manchesters, owing to the bank on the left shutting off the view. The firing line swept on out of the depression to about 250 yards from the redoubt. In front was now an open glacis-like slope, and in crossing this the Regiment suffered most severely, the enfilade fire from the right being particularly deadly.

Some survivors of the firing line, under Captain Wynter, penetrated into the trenches of the redoubt. Captain Wynter had already been wounded once. Captain Heath was wounded on this slope in front of the redoubt, and Lieutenant Scobie was killed here. Some men of No. 2 Company led by Indian officers, attempted to advance up a trench leading from some water holes in front of the redoubt into it, but they were bombed back, all our bombers by this time being casualties. No. 1 Company (reserve) under Captain Gordon and Lieutenant Milligan, now tried to rush up the slope. Both were wounded, Lieutenant Milligan was eventually missing and presumed killed. He was never heard of again. A few of No. 1 Company got into the redoubt. None of them returned or were ever heard of again. On the left some Manchesters had got into the redoubt, after having suffered heavily in the advance, but they were counter-attacked by the Turks and driven back. The 59th rallied at the depression in front of the redoubt and got back Captain Wynter, who had been wounded twice more. A stand was made here and much loss inflicted on the Turkish counter-attack. After this a general retirement took place, covered by the 47th Sikhs, who had previously taken up a position with this object in view. The 59th Rifles now ably rallied by Captain R. D. Inskip, M.C., and reorganized by their Indian officers (practically all British officers were killed or wounded), retired steadily and in excellent order. The machine-gun section under Subadar-Major Perbhat Chand, M.C., covered the Regiment from a well-selected position on the right flank. This section suffered severely from hostile artillery fire, one gun being knocked out by a direct hit. Subadar-Major Perbhat Chand and half the gun teams were wounded, but the gun commanders kept the remaining guns in action in a most plucky and praiseworthy manner.

The enemy trenches were flush with the ground with no parapets showing and so were quite invisible. Their machine guns seemed to be in emplacements well forward of the position, so that they were able to cover their line with oblique and enfilade fire. Of the three other brigades which were said to be taking part in the attack, nothing was seen or heard. From reports received from enemy officer prisoners captured some five weeks later, it appeared that if one more brigade had joined in they would have gone back, and orders had already been issued for a Turkish retirement !

At 9 p.m., when the Brigade had somewhat re-formed, piquets were put out by Nos. 2, 3, and 4 Companies with No. 1 in support, total strength some 200 men. All ranks were completely exhausted and suffering greatly from want of water.

Casualties for the day were :—

British Officers.—Lieutenant J. A. M. Scobie, killed ; Lieut.-Colonel T. L. Leeds, Captains Gordon, Heath and Wynter, wounded ; and 2/Lieutenant J. R. Milligan, missing.

Indian Officers.—Subadar Rukam Din (52nd), killed ; Subadar-Major Perbhat Chand, Subadars Sahib Singh, Bishen Singh, Tola Singh, Jemadars Amir Khan and Ghamai Khan, wounded.

Indian Other Ranks.—Killed, 43 ; wounded, 105 ; missing, 10.

Early next morning orders were received for the force to retire to Wadi Camp. Colonel Leeds, who was lying wounded in an advanced dressing station on the night of 8th-9th, was informed by the Medical Officer that orders had been received to abandon wounded and retire. The Medical Officer asked Colonel Leeds whether he could arrange to get himself moved and also Subadar-Major Perbhat Chand and Jemadar Ghamai Khan, who were lying beside Colonel Leeds. Fortunately 59th headquarters were near at hand, and men were produced who carried all three nearly a mile in the dark to the field ambulance. Again on the morning of the 9th there was a discussion as to whether wounded should be left behind or taken away, but, fortunately, for the 59th especially, it was decided to take all wounded away, and a convoy started at 10 a.m. The rearguard was followed

Sketch Map showing the attack on Dujailah Redoubt

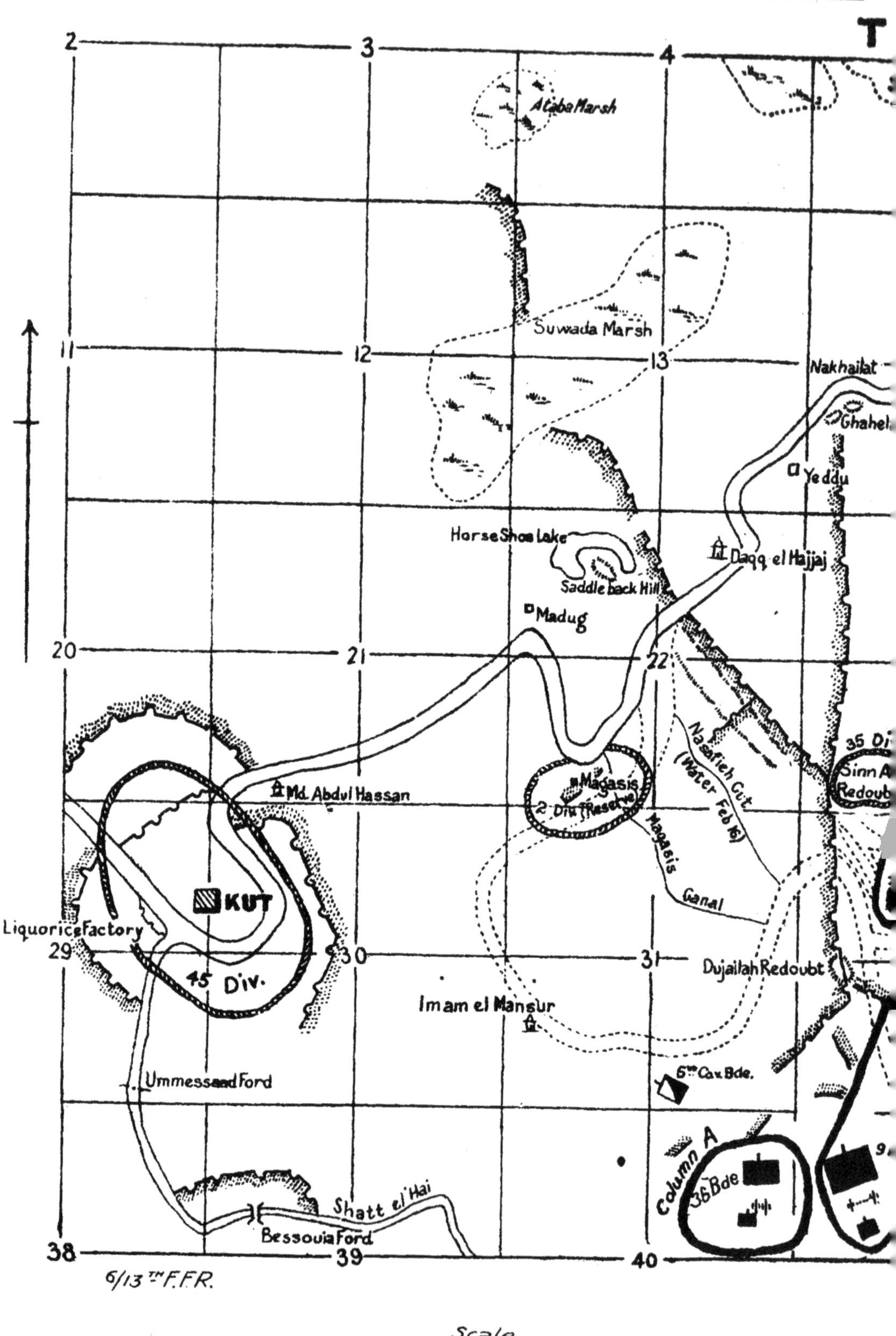

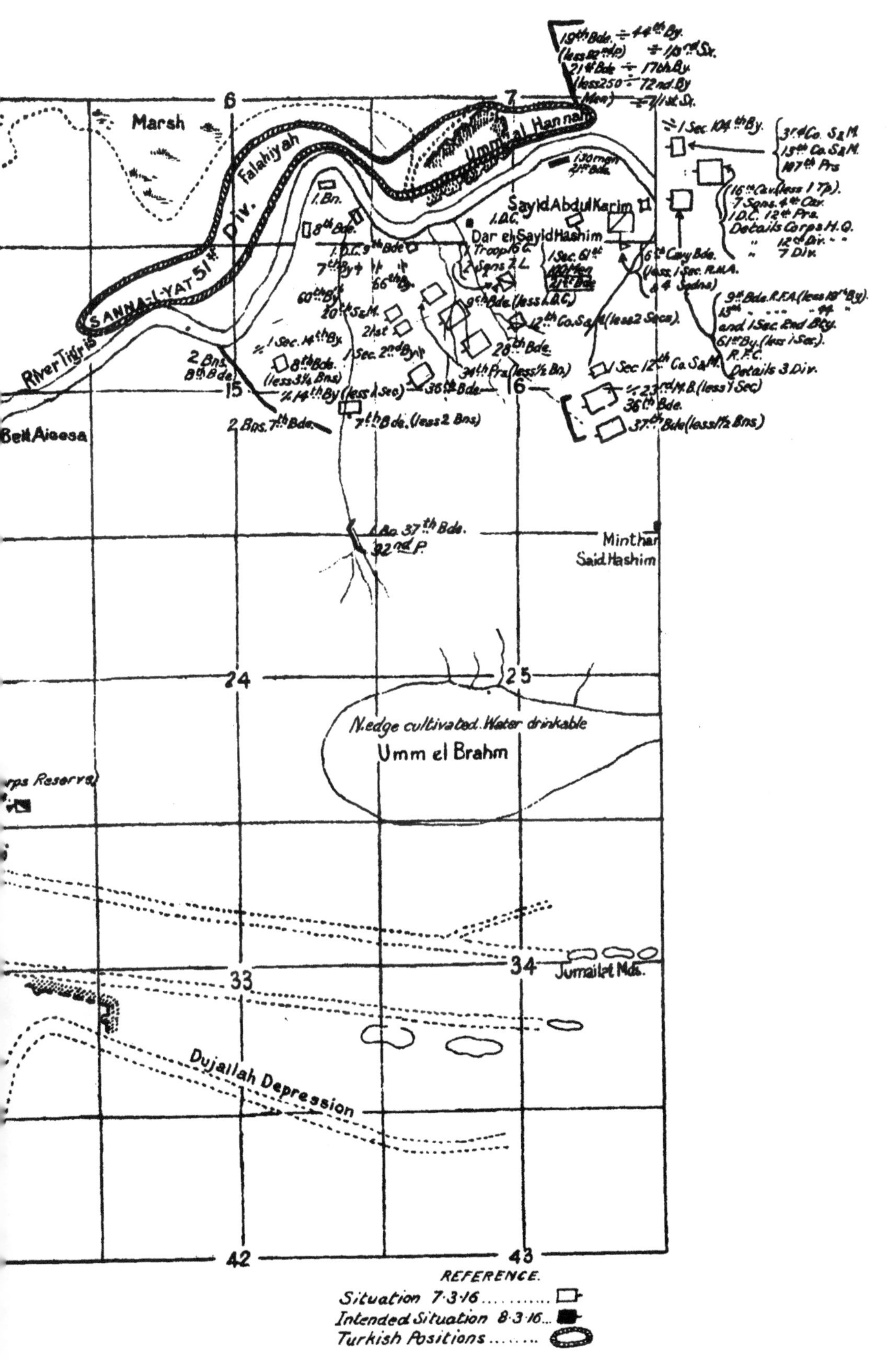
Marsh
Falahiyah
SANNA-I-YAT 51st Div.
Umm al Hannah
River Tigris
Sayid Abdul Karim
Dar el Sayid Hashim
Beit Aiessa
Minthar
Said Hashim
N. edge cultivated. Water drinkable
Umm el Brahm
Jumailat Mds.
Dujailah Depression
6
7
15
16
24
25
33
34
42
43
REFERENCE.
Situation 7·3·16
Intended Situation 8·3·16
Turkish Positions

up and shelled, and the convoy of wounded was bombed by a hostile 'plane, causing more casualties amongst those already wounded. The retirement took all day, and great hardship was endured owing to the entire lack of water. Camp was reached at 8.15 p.m. after a most exhausting day. Captain R. D. Inskip was now in command of the Regiment. Many congratulatory messages were received by the Regiment, and the 59th were subsequently mentioned in Sir Percy Lake's despatch, as having particularly distinguished themselves at Dujailah, and in the action which took place on April 18th at Beit-Aiessa.

On March 11th information was received that the 7th Brigade had retaken the Abu Roman position, which had been evacuated by our troops, and that 2 officers and 50 prisoners had been taken. The G.O.C. 3rd Division came and congratulated the Regiment on its gallant conduct and good work on March 8th. Captain H. G. Turner, 106th Pioneers, arrived on the evening of March 12th and took over command. He left again on April 1st, when Captain Inskip resumed command. On the 14th Major-General Keary, C.B., D.S.O., inspected the Brigade, and congratulated everyone on the way the Brigade assaulted the Dujailah Redoubt on the 8th.

On March 25th the 8th Brigade relieved the 7th Brigade in the trenches. The 59th were in reserve at Thorny Nullah. The Tigris burst its banks by Mason's Mound, and ground between the enemy and the 8th Brigade was flooded. The 47th and 59th were flooded out of their trenches, and had to dig a new line on carefully sited ground. Information was received on the 29th that Captain J. R. Wynter had been awarded the D.S.O. (immediate award) for conspicuous gallantry at Dujailah.

On April 2nd Brigadier-General S. M. Edwardes, D.S.O., took over command of the 8th Brigade. The 59th moved to Sandy Ridges on April 5th. The following officers joined for duty on April 4th:—Captain W. H. Millar, 74th Punjabis, 2/Lieutenants Chapman, Dudley and Morrison, the last-named on return from a Machine-Gun Course.

On April 5th the recently arrived 13th Division assaulted the Hanna position and took it in one hour. Two or three hundred Turks retreated across the 59th front, and were severely punished by machine-gun and rifle fire. The 13th Division advance continued next day, and as much support as possible was given by fire from the right bank across the river, but the enemy did not present suitable targets, except in the afternoon, when the Regiment took full advantage of them. The 13th Division captured the Fallahiyeh position at 7 p.m. Next day the Regiment joined the Brigade at Abu Roman.

At 5.45 a.m. on the 6th the attack on the Sanna-i-Yat position on the opposite bank by the 7th Division commenced. The Hanna and Fallahiyeh positions captured by the 13th Division on the previous day had not been strongly held; Sanna-i-Yat was the main position, which the Turk intended to hold at all costs. The 7th Division captured about 400 yards of front-line trench, but at great cost. A further attack was ordered at dusk, but was postponed on account of a sudden rise in the river. The 8th Brigade moved forward to a position on the right bank opposite Sanna-i-Yat. The 59th dug in on a 500-yard front, with Manchesters on the right and 37th Brigade on the left. Sanna-i-Yat was heavily bombarded next morning at 7 a.m., and again two hours later when enemy reinforcements were seen coming up. Bombardment was continued in the evening. At 4.45 a.m. on April 9th the 7th and 13th Divisions assaulted the Sanna-i-Yat position, but by 8 a.m. news was received that the attack had failed.

On April 12th the 59th received orders to advance, drive in the enemy's piquet line, and advance on to his main position (the ground had been reconnoitred by several officers' patrols during the last few days). At 3 p.m. the line advanced and waded through the flood to the enemy's piquet line, which was found to be abandoned. A line was established here and held till dark. In the meantime touch was lost on both flanks, as units there were unable to advance on account of the floods. This was reported to the Brigade, who ordered a retirement until touch could be re-established. This was done, and a line occupied some 700 yards in rear on the edge of the marsh. In the advance 2/Lieutenant Chapman was slightly wounded, and Subadar Bukhan Singh and Jemadar Hira Singh were killed, and 30 Indian other ranks were wounded.

On April 14th Lieutenant-Colonel T. L. Leeds rejoined from hospital, and a draft arrived consisting of 2 Indian officers and 143 other ranks. On April 15th the 7th and 9th Brigades brought off a successful attack, and the 8th Brigade moved forward to conform with the new line. At noon the 59th were ordered to rush and hold the enemy piquet line. Two parties of 30, each under Indian officers, were detailed for this operation. These parties advanced at 1.30 p.m., and were heavily fired on, but were completely successful. Lieutenant Culverwell went up to superintend the consolidation of the new line, but was immediately severely wounded in the thigh. Subadar Abdullah Shah and 14 other ranks were also wounded in this operation. Captain Stirling then went forward to superintend consolidation.

On the morning of April 17th the 7th and 9th Brigades brought off a most successful attack on the Beit-Aiessa position, gaining all their objectives. The 8th Brigade were in support, the 59th holding some 1,800 yards of front with only about 400 men. The 47th, on the right, also had a very extended front. Some of the Manchesters were on the left, and then the Brigade flank seemed to be more or less in the air. During the afternoon the Commanding Officer had observed Turkish troops massing and moving across the front from left to right; about the equivalent of a strong division had been seen. This was reported to Brigade headquarters, and the Brigade Major himself saw the movement from Battalion headquarters still going on at about three miles away. This was reported to Corps and Division. At 8 p.m. the storm broke in the shape of a strong and well-organized counter-attack on the 7th and 9th Brigades on the right. The 7th Brigade front was broken through by the enemy, but for some unaccountable reason he did not exploit his success, but switched his attack on to the portion of the line held by the 47th and 59th. It was a bright moonlight night, and the enemy could, fortunately, clearly be seen. He came on time after time, getting to within 15 and 20 yards of the front line, but was shattered and beaten back by fire on each occasion. There were several gaps in our line where no trenches existed. One of these was 400 yards long. Fortunately, the enemy never found or penetrated at these points. Ammunition began to run short, and the 47th were also running short of ammunition. Every available rifle was in the firing line, headquarters being reduced to the barest possible minimum.

Subadar Gauri Charan then organized a carrying party, assisted by Havildar Qalandar Khan. This party had to cross a perfectly level space of over 1,000 yards from Battalion to Brigade headquarters to bring up the much needed ammunition. Repeated journeys were made by this party, and enough ammunition was brought up to supply everyone and also to form a substantial reserve of small arms ammunition and bombs as well. Subadar Gauri Charan and his party displayed the greatest gallantry that night, and undoubtedly saved the situation. The only artillery support was obtained from the 23rd Mountain Battery, whose guns were worn out and whose ammunition soon ran out.

At one point of the line there was a trench running towards the enemy. Havildar Sohnu and a party of bombers held this trench, up which the Turks sent constant bombing attacks, but Havildar Sohnu maintained his position successfully throughout the night, and did great execution among the enemy. The trench was found to be piled with dead in the morning. A machine-gun of the Regiment posted on the left flank was able to enfilade the front line, and did great execution and helped considerably in breaking up the many attacks, which were kept up throughout the night.

At about 2 a.m. it became clear that the enemy was nicely held. There was plenty of ammunition, and there was no further cause for anxiety. As soon as it was light enough to observe, the artillery opened intense fire, and the enemy broke and ran. In the morning it was calculated that there were at least 2,000 Turkish dead in front of the 47th and 59th lines. There were Arab troops in this counter-attack, and it was found to be dangerous work trying to collect or help wounded, as wounded Arabs frequently shot at rescue parties. The G.O.C. sent the following message early in the morning :—" Express to all ranks G.O.C.'s appreciation of their real gallantry " ; and later he came in person and congratulated the Regiment on the splendid work it had done during the night. The Regiment was also mentioned in a special Order of the Day and in Sir Percy Lake's despatch.

The casualties for the night were—Killed, Captain W. H. Millar and 6 Indian other ranks ; wounded, S.A.S. Chur Singh, Subadar Amin Khan, and 24 Indian other ranks.

The Regiment was relieved at 2 p.m. on the 18th, and went into billet trenches at Rhodes Piquet, having been marching and fighting for forty-eight hours on end.

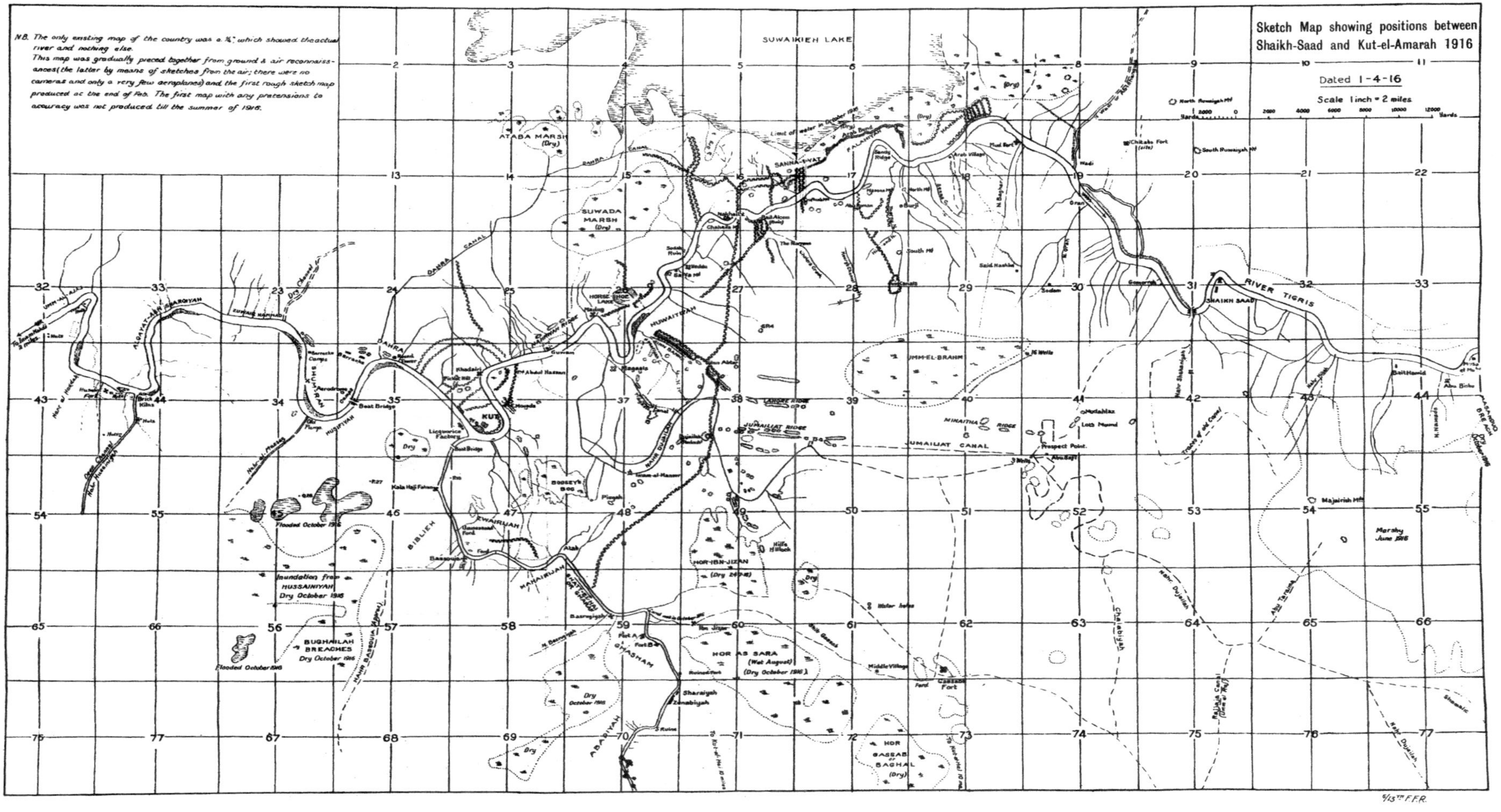

Sketch Map showing positions between
Shaikh-Saad and Kut-el-Amarah 1916
Dated 1-4-16
Scale 1 inch = 2 miles
N.B. The only existing map of the country was a ¼" which showed the actual river and nothing else.
This map was gradually pieced together from ground & air reconnaissances (the latter by means of sketches from the air; there were no cameras and only a very few aeroplanes) and the first rough sketch map produced at the end of Feb. The first map with any pretensions to accuracy was not produced till the summer of 1916.
SUWAIKIEH LAKE
ATABA MARSH (Dry)
SUWADA MARSH (Dry)
SANNA-I-YAT
RIVER TIGRIS
SHAIKH SAAD
UMM-EL-BRAHM
JUMAILIAT RIDGE
JUMAILIAT CANAL
MIKAITHA RIDGE
HUWAIYAH
KUT
SHUMRAN
DAHRA
DAHRA CANAL
BIBLIEH
KHADAIRI
HORSE-SHOE LAKE
HOR-IBN-JIZAN
HOR AS SARA
HOR GASSAB or BAGHAL (Dry)
BUGHAILAH BREACHES
Dry October 1916
Inundation from HUSSAINIYAH Dry October 1916
Flooded October 1916
Chalabiyah
Nahr Dujailah
Prospect Point
Majairish Mᵈ
Marshy June 1916
To Kut-el-Hai 10 miles
5/13ᵗʰ F.F.R.

N.B. The only existing map of the country was a ¼" which showed the actual river and nothing else.
This map was gradually pieced together from ground & air reconnaissances (the latter by means of sketches from the air; there were no cameras and only a very few aeroplanes) and the first rough sketch map produced at the end of Feb. The first map with any pretensions to accuracy was not produced till the summer of 1916.

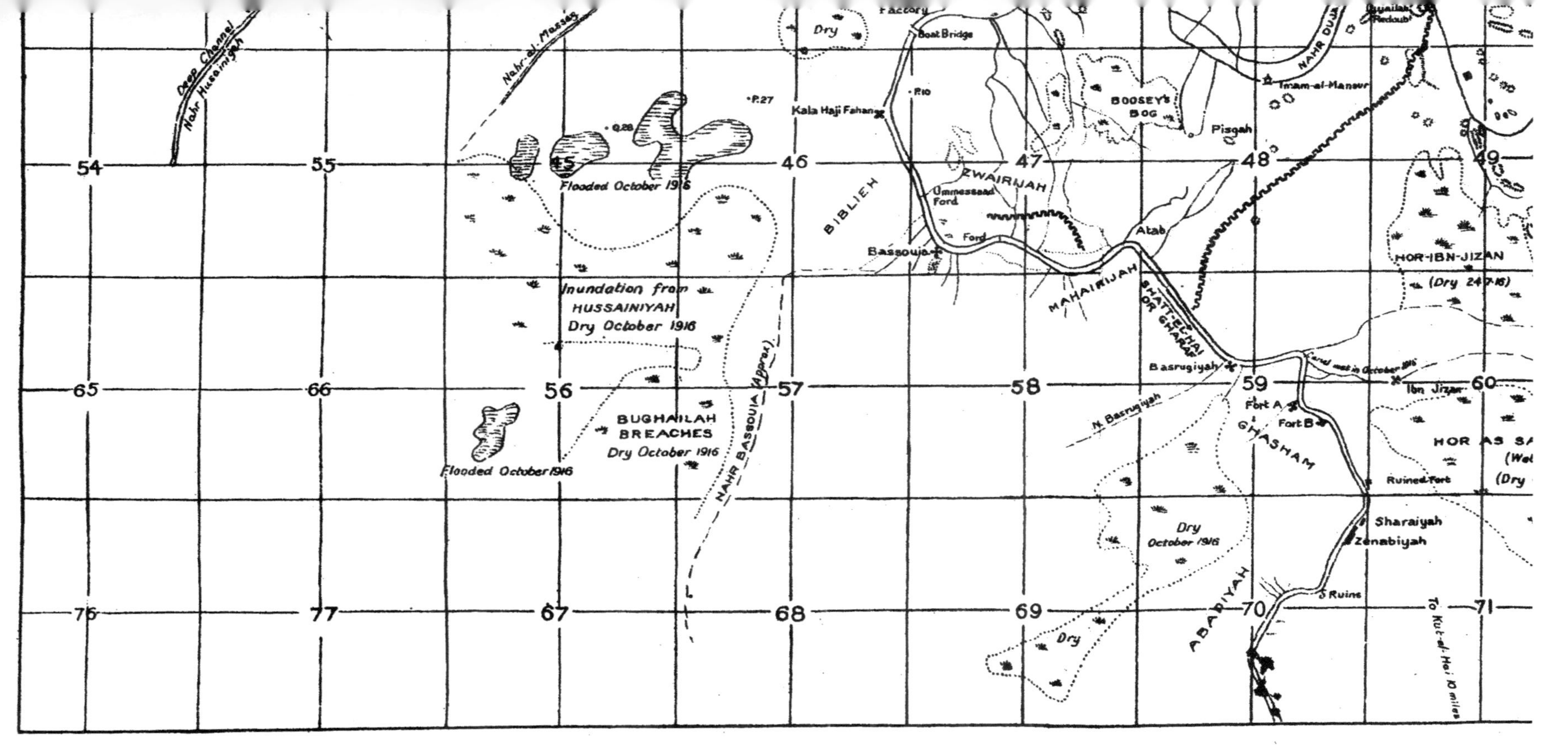
Deep Channel
Nahr Husainiyah
Nahr-al-Masses
Dry
Factory
Boat Bridge
P.27
Q.28
Kala Haji Fahan
P.10
BOOSEY'S BOG
Imam-al-Mansur
NAHR DUJ
Dujailah Redoubt
Pisgah
Flooded October 1916
ZWAIRIJAH
Ummessead Ford
BIBLIEH
Ford
Bassouja
Atab
HOR-IBN-JIZAN
(Dry 24.7.16)
MAHAIRIJAH
SHATT-EL-HAI OR GHARAF
Inundation from HUSSAINIYAH Dry October 1916
Basrugiyah
Canal met in October 1916
Ibn Jizan
NAHR BASSOUJA (Approx.)
BUGHAILAH BREACHES Dry October 1916
N. Basrugiyah
Fort A
Fort B
GHASHAM
Flooded October 1916
HOR AS SA
(Wet
(Dry
Ruined Fort
Dry October 1916
Sharaiyah
Zenabiyah
ABADIYAH
Ruins
To Kut-el-Hai 10 miles
Dry
54
55
45
46
47
48
49
65
66
56
57
58
59
60
76
77
67
68
69
70
71

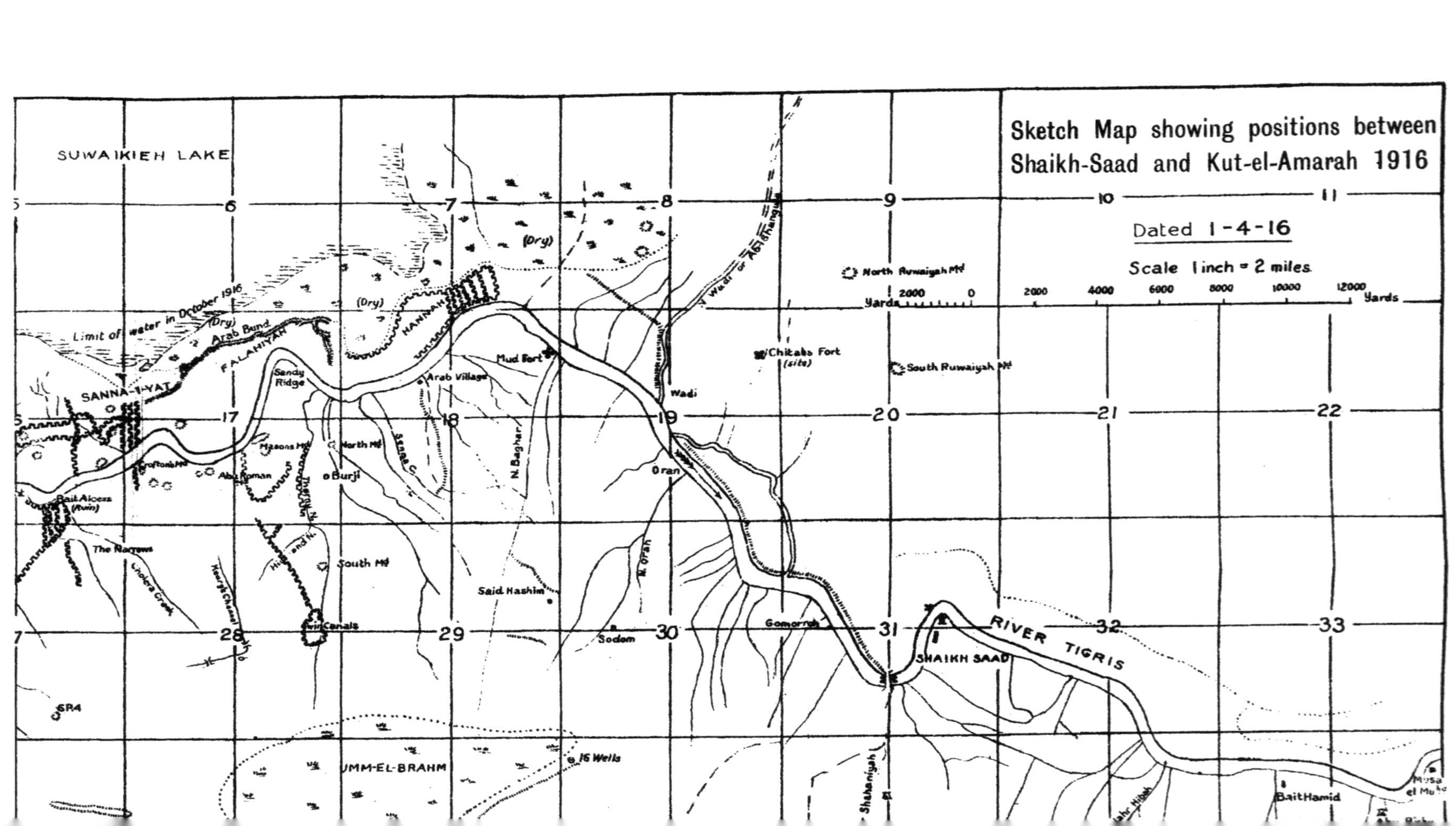
Sketch Map showing positions between Shaikh-Saad and Kut-el-Amarah 1916
Dated 1-4-16
Scale 1 inch = 2 miles
Yards 2000 0 2000 4000 6000 8000 10000 12000 Yards
SUWAIKIEH LAKE
Limit of water in October 1916
(Dry)
Arab Bund
FALAHIYAH
HANNAH
SANNA-I-YAT
Sandy Ridge
Mud Fort
Arab Village
Wadi
Chitabs Fort (site)
North Ruwaiyah Md
South Ruwaiyah Md
Masons Md
North Md
South Md
Burji
Abu Roman
Sanna C.
N. Bagher
Oran
N. Oran
Said Hashim
Sodom
Gomorrah
SHAIKH SAAD
RIVER TIGRIS
Bait Aiessa (Ruin)
The Narrows
Cholera Creek
Keary's Channel
Canals
SP4
UMM-EL-BRAHM
16 Wells
Shahaniyah
Bait Hamid

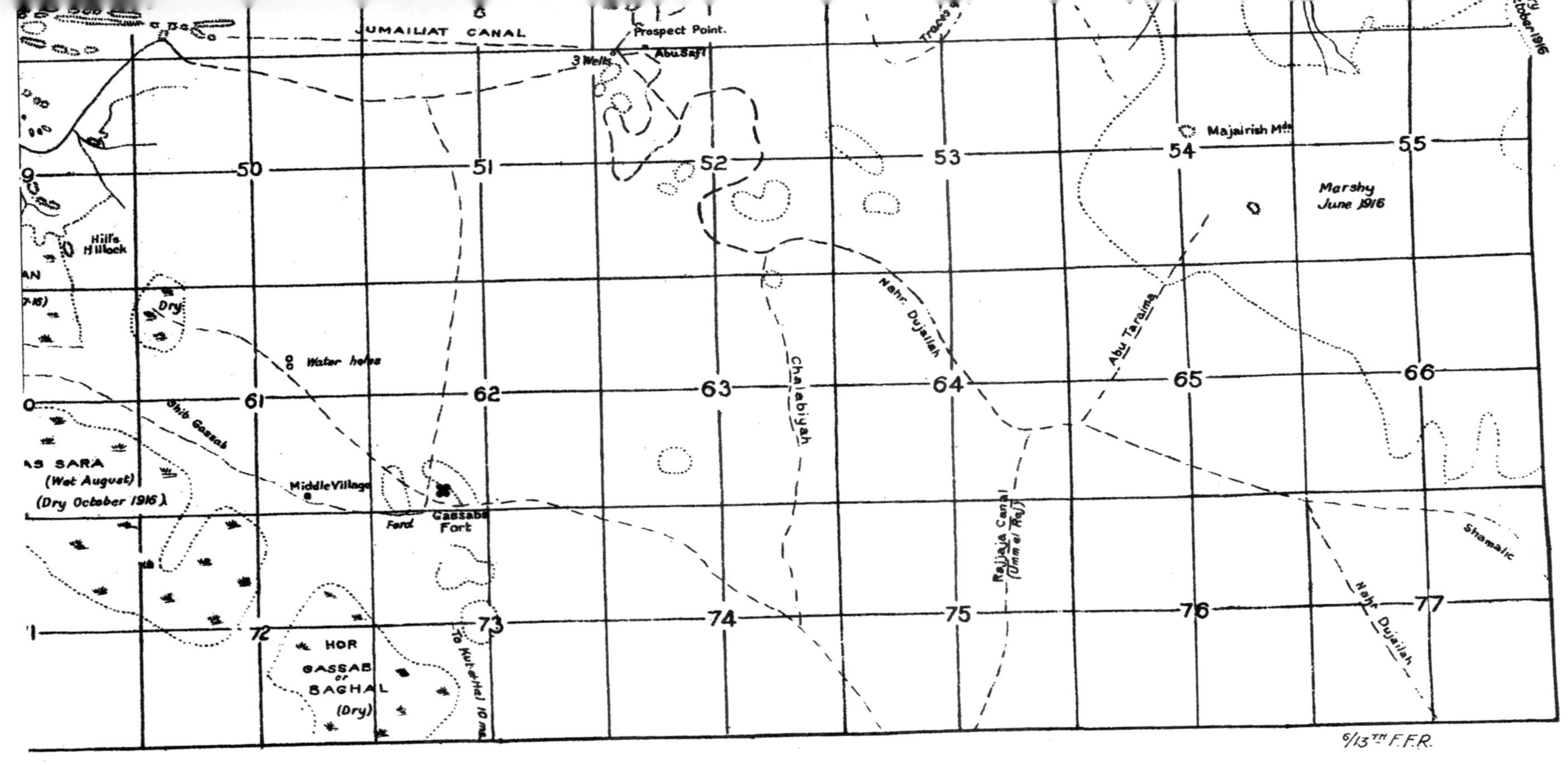

JUMAILIAT CANAL
Prospect Point.
3 Wells
Abu Safi
Majairish Mds
Marshy
June 1916
Hill's Hillock
Dry
Water holes
Shib Gassab
Chalabiyah
Nahr. Dujailah
Abu Taraima
Rajjaja Canal
(Umm el Raj)
Shamalic
AS SARA
(Wet August)
(Dry October 1916)
Middle Village
Ford
Gassaba Fort
To Kut-el-Hai 10 ms.
HOR
GASSAB
or
BAGHAL
(Dry)
50
51
52
53
54
55
61
62
63
64
65
66
72
73
74
75
76
77
6/13TH F.F.R.

Final Attempts to relieve Kut.

On the 22nd the 7th Division delivered one more attack on the Sanna-i-Yat position, but were again beaten back. Two days later the 9th Brigade brought off a successful attack, and the 3rd Division was able to advance a couple of miles to a position facing the Jumailat Ridge, with Dujailah Redoubt on the right. The Regiment held a piquet line of about 800 yards' frontage. On April 25th a steamer, the *Julnar*, manned by volunteers from the Royal Navy, tried to get through to Kut with rations, but the advent of the steamer was well known to the Turks (spies at Amara had seen her loading there and knew what was intended), who put a steel hawser across the river and trained their guns on it. The Navy made a most gallant attempt, but the steamer was captured, and the crew suffered very heavy casualties. All attempts to relieve Kut had now failed, and news was received that the garrison had been forced to surrender on April 29th.

On May 2nd a truce was arranged for an exchange of Kut sick and wounded. Detachments of the 3rd Division were inspected by Sir Percy Lake, who made special mention of the good work done by the 8th Brigade in France and Mesopotamia.

On May 9th news of the awards for gallantry in action on April 17th-18th was received.

On May 20th the 3rd Division advanced towards the Dujailah and Sinn Abtar Redoubts. The advance though unopposed was most trying, on account of intense heat, lack of water, and delays caused by the transport, to which the 59th were acting as escort. Three men of the Regiment were sent to field ambulance with heat-stroke. The road was littered with men of the leading brigades, who had fallen out. The G.O.C. Division complimented the Regiment on its splendid marching abilities. The enemy made no attempt to oppose the advance of the Division. The 3rd Division remained in this new position extending from the Magasis Canal to Dujailah Redoubt, until relieved by the 14th Division on June 18th. The Regiment moved to Abu Roman on June 28th, and thence to Thorny Nullah on July 1st, and remained there until August 21st. October was spent at Sinn Abtar. Nothing of interest happened during the rest of the year. Two drafts joined between June and December.

Six British officers, two Indian officers, and 14 Indian other ranks were mentioned in General Sir Percy Lake's despatch for distinguished work between January and April, 1916.

1917. In the meantime General Maude had been appointed to command the Mesopotamia Expeditionary Force. Under his command the most wonderful and beneficial change occurred. Troops were well fed and cared for, trained, and brought to a high state of efficiency. Many steamers were sent to Mesopotamia, a metre gauge railway was constructed from Basra to Amara, and all transport and supply arrangements were made as good and efficient as possible. How well laid General Maude's plans were, and how efficient all arrangements in Mesopotamia became under his able guidance and supervision, the events of 1917 most amply proved. Every operation undertaken was crowned with success, and the Army in Mesopotamia had complete confidence in their wonderfully able commander.

January 1st, 1917, found the Regiment in trenches which were being gradually advanced with a view to clearing the right bank of the Tigris of the enemy, and the ultimate capture of Kut, and an advanced line was constructed by January 8th, from which an attack was to be delivered on the following day. Several casualties occurred during the construction of this position on January 2nd.

At 9 a.m. on the 9th, the Manchesters advanced from their position immediately after a heavy bombardment, and covered by an artillery barrage. The advance was so well timed, that they reached the enemy trenches before the Turks realized that the barrage had lifted, and the enemy were found still lying at the bottom of their trenches. No. 1 Company followed up in support of the Manchesters. This company was informed that reinforcements, S.A.A. and bombs were required on their right, so they bore off in that direction, and got into the Turk trenches, and successfully bombed the enemy back until they reached a nullah, converted into a trench, with island traverses, bearing off to the left. Half the company went along the main trench while the other half bombed up, and secured the head of the nullah. The attack along the main trench met with such strenuous resistance that it resolved itself into a bombing duel, until Sepoy Kahn Singh, on his own initiative,

RESTING. 1916.

CAPTURES FROM THE TURKS, 1917.

1.—Amara, 1916. Captain J. A. M. Scobie, M.C. (killed 8/3/16), in foreground.

2.—Ctesiphon Arch.

3.—The Citadel, Baghdad. Scaffold in centre.

4.—The *Firefly*, after recapture from the Turks.

climbed out of the trench with his Lewis gun, which he shouldered and opened fire on the enemy in the trench below. He drove the enemy out of the trench, and down a second nullah, which led into the first one. Sepoy Kahn Singh continued to fire his Lewis gun at the enemy until a bomb exploded on and broke the stock of his gun, severely wounding Kahn Singh in the shoulder. For this splendid act of gallantry he was given an immediate award of the Indian Order of Merit (2nd Class). The Turks in the second nullah were superior in numbers to the half-company which had fought their way along the main trench, and our men suffered many casualties. No. 2 Company came up at this moment and continued the attack to the right. Captain Cooper, commanding No. 2 Company, was at this point severely wounded. At this stage a mixed body of troops was driven back from the right flank on top of the men of Nos. 1 and 2 Companies, and caused much confusion and crowding in the trench, the Turks counter-attacked across the open and down the nullah, and someone in the aforesaid mixed troops started the order "about turn," which developed into "retire." The men of Nos. 1 and 2 Companies in this portion of the trench, were completely disorganized by this mob, and many fell back with them to the forward position. Here they were reorganized, with two platoons of No. 4 Company as support, and formed into a fresh wave and attacked once more, but as they were advancing, the mist which had been very thick all the morning, lifted, and the enemy seized their opportunity and opened a heavy fire on the attack. Two attempts were made to regain the position, but failed owing to the heavy casualties inflicted, and the attack had to be abandoned until adequate artillery support could be arranged. At 1 p.m. No. 4 Company was sent up to the Turkish trenches on the left and No. 3 followed at 8 p.m., the line having been consolidated in the meantime. At 11.30 p.m. the Regiment was collected in the trench from which the original attack had been delivered. The casualties for the day were as follows :—

Killed—Acting Lieut.-Colonel H. F. D. Stirling, M.C.
Captain J. H. Manley.
Subadar Kishen Singh.
Jemadar Mit Singh.
28 Indian other ranks.

Died of wounds—Lieutenant R. Davis.
Subadar Amin Khan.
Jemadar Abdul Wahab, I.O.M.
Missing—7 Indian other ranks.
Wounded—Captain H. L. Cooper.
Captain R. H. Burne.
Subadar Bahadur Shah.
70 Indian other ranks.
Total casualties—115.

The following British officers were present in this engagement:—

Captain and A./Lieut.-Colonel H. F. D. Stirling (killed).
Lieut. H. L. Cooper (wounded).
Lieut. W. A. H. Young.
Lieut. R. H. Burne (wounded).
Lieut. J. D. Twinberrow.
Lieut. R. Davis (died of wounds).
Capt. A. H. Burn.
Lieut. E. J. K. Garthwaite.
Lieut. R. S. Dudley.
2/Lieut. J. H. Manley (killed).
2/Lieut. A. Morrison.

On January 10th the Regiment went out of the line until the 13th. A few casualties occurred between the 13th and 18th, on which date all preparations were made to drive the Turk off the right bank, but during the night 59th patrols discovered that the enemy had abandoned their position. The Turkish trenches were occupied by the Regiment, and the result of this excellent patrol work was reported. This information caused the 9th Brigade to investigate the trenches in front of them. These were also found to have been abandoned, so that the bombardment of an evacuated position was saved.

At the beginning of February there were successful attacks on both banks of the Hai. The Turks opened heavy shell fire on the 3rd, and continued it in the evening. The Regiment suffered over twenty casualties from this. At 12.45 a.m. on the 4th, 59th patrols discovered the trenches opposite evacuated and so occupied them. At noon on February 9th the top of the minaret at Kut was knocked off by a direct hit from one of our guns. The Sanna-i-Yat position on the left bank was attacked by the 7th Division on the 17th, but without success. Lieut.-Colonel T. L. Leeds rejoined from field ambulance on the 19th and resumed command. Lieutenants Morrison, Malden and Clapham reported their arrival, the two last mentioned on first joining.

1.—**Meeting of British and Russian Forces at Kizil Robat, March, 1917. Brig.-Gen. S. M. Edwardes and Staff of 8th Jullundur Bde.**
2.—**Marching up the Tigris, 1916. (*Left to right*): Col. Leeds, Capts. Luck, Inskip, Baker, Scobie, Lieuts. Stuart-Prince and Bickford.**
3.—**Crossing the Diala, 1917.**
4.—**Baquba. Sub.-Major Perbhat Chand, M.C., in foreground.**

LIEUT.-COLONEL T. L. LEEDS, C.M.G., D.S.O.

Who served with the Regiment through the War and commanded from March, 1915, until February, 1921.

On the 20th the 47th Sikhs and the 59th Rifles had moved over to the left bank and become Corps reserve at Fallahiyeh, ready to move at half-an-hour's notice. They were joined by the rest of the Brigade two days later. The 8th Brigade followed up the 7th Division on the 23rd, and marched from Sanna-i-Yat to the Sawada position. After putting out outposts orders were received to continue the advance, which went on uninterruptedly until Baghdad was reached on March 14th. Kut and its defences were passed on February 25th, and Ctesiphon was passed on March 10th. During the latter part of the march blinding dust storms were experienced.

On the 22nd the 7th Division successfully assaulted the Sanna-i-Yat position at 10 a.m.; the 28th Brigade delivered a second attack in the afternoon on the right bank, and this too succeeded. The Turk was now really on the run, and he did not stop until he reached Samarra, seventy-five miles beyond Baghdad.

On March 18th information was received that a strong force of the enemy was holding Baquba and Buhriz, on the left bank of the Diala, and orders were received for the Manchesters and 59th Rifles, under Lieut.-Colonel T. L. Leeds, to cross the Diala, and seize the above-named places. Both regiments were successfully ferried across the Diala, about 100 yards wide at this spot, with a strong current, and Buhriz was captured after slight opposition, and a platoon and two machine guns under Lieutenant Garthwaite were left there. At 9.30 a.m. Baquba was reached and found to be unoccupied by the enemy. People here seemed very pleased to see British troops, and a large supply of excellent oranges was obtained. Here good billets were obtained in large, well-built houses. Piquets were put out round the town, and patrols sent out. Several Turks, chiefly sick and wounded, were collected and made prisoners. The platoon at Buhriz was relieved, and rejoined the Regiment.

On March 20th the Turks were reported to be marching on Baquba, and the 8th Brigade were ordered to move out and take up a defensive position ten miles beyond Baquba. No enemy were encountered, and a further advance was ordered. The cavalry, ahead of the Brigade, came up with the Turks near Jilali, and drove them back. The enemy continued retiring, blowing up bridges over canals, until he reached the

Jabal Hamrin range of hills. On the 23rd the 59th Rifles supported by artillery, and one section Machine Gun Company, all under Lieut.-Colonel T. L. Leeds, D.S.O., were ordered to advance on the Jabal Hamrin position at 1 p.m. Orders were not to become seriously engaged but to ascertain the enemy's strength. Major Burn commanding the point of the advanced guard, sent back at 2.20 p.m. to say that he had halted behind the bund of a narrow canal, and was pushing out scouts. He was being heavily shelled and sniped, and seven enemy guns had been located on the left and right of the road. The Officer Commanding advanced guard sent up all available artillery, who galloped across the open, one entire team being knocked out by a direct hit, and came into action, not far behind the narrow canal. All the 59th were pushed up too, and headquarters established thereon the Nahrunia Canal by 3.45 p.m. The Beled Ruz Canal was found to be fifty or sixty yards wide, deep and swift. The enemy had sniping posts on the far side, and the bridge across was completely destroyed. This was reported to the Brigade. Casualties during the day were :—Jemadar Bhan Singh and 21 other ranks wounded, 5 other ranks killed. That night information was received that pontoons were being sent up and that an attack would be made. A Sapper Officer inspected the bridge, and decided that no bridge could be made in time, so the attack was postponed, and a strong piquet left at the bridge-head. This piquet was much worried by snipers on the far bank, who were eventually located and dealt with.

On the morning of the 25th the 9th Brigade assaulted the Jabal Hamrin position, crossing the canal about two miles to the east, and attacking the enemy's left flank. They were supported by artillery fire from the 8th Brigade, and were at first successful, but later, in the hills on the right front of the 8th Brigade position, met superior enemy forces who outflanked them, and had to withdraw after suffering heavy casualties. Their retirement was covered by the Manchesters and the 124th Baluchis. On this day the 59th lost Jemadar Gul Sahib, mortally wounded, 3 other ranks killed, and 7 wounded. The bridge across the Beled Ruz was completed at 5.15 a.m. on the 26th. During the day the bridge piquet was heavily sniped, and had eight men wounded. That night patrols from the Regiment investigated the reeds on the opposite bank, and found no one there. At 7.50 a.m. on the 27th patrols went up a nullah opposite, which had been occupied by the enemy and found it vacated. It had been strongly held as was shown by the large number of empty cases, and snipers' posts. This nullah was then occupied by three platoons of No. 4 Company, who were promptly shelled, but fortunately no one was hit. The line was again advanced about 1,000 yards at 10.30. On the 29th air reports stated that the Jabal Hamrin had been evacuated, but patrols sent out were promptly fired on. In the afternoon strong patrols of the Regiment penetrated into the foothills unopposed, and No. 2 Company under Lieutenant Roseveare got on to the outer crest of the Jabal Hamrin. They were recalled by the order of the G.O.C. Division at about 5 p.m. Lieutenant Roseveare reported that patrols sent on from his position had been fired on by machine guns about 900 yards on farther into the hills, where he had seen small bodies of Turks.

On March 30th the 59th Rifles with a battery R.F.A. and a section of machine guns carried out a reconnaissance in force. It was discovered that the enemy was holding a strong piquet line about 1,000 yards beyond the crest. The situation appeared to be that these piquets could be driven back, but this would entail heavy fighting, and the orders were against getting heavily engaged. A report was sent back, and the Division ordered a withdrawal. It took over two hours to break off the fight, and get the Regiment back on to the outer crest.

On the following day the 47th Sikhs reconnoitred the Jabal Hamrin and found that the Turks had left during the night. The 47th remained in the hills that night, and the rest of the Brigade marched to Kizil Robat on the following day. The Turks had crossed the Diala at this spot by the time the 8th Brigade arrived. A squadron of Cossacks rode into camp from Khanikin just after noon on April 2nd, and excited much interest. The Brigade then marched back to Baghdad, arriving there on April 7th, and remained there until the 16th, getting a much-needed rest, and also some summer clothing.

The 8th Brigade marched up the line again on April 17th, and arrived at 7th Divisional Headquarters early on the morning of April 21st and became Corps reserve. At 5 a.m. gun-fire flashes, the beginning of the Battle of Istabulat, could be seen. The Turks were still strongly entrenched here, and the Regiment was under orders to be ready to move at half-an-hour's notice. In the evening the Turks gave half-an-hour's furious bombardment, which by now had come to be recognized as the prelude to his retiring after darkness. The 7th Division occupied the Istabulat trenches on the 22nd, and the enemy retired to a hastily dug line about two miles north of Istabulat. The 28th Brigade, supported by the 9th Brigade (the 8th Brigade and two batteries of Artillery were guarding our left flank) attacked the Turk in his new position and drove him out, but suffered severely in doing so. On the 23rd the 8th Brigade advanced through the 7th Division and took up the pursuit. Cassel's Cavalry, who had been sent to make a wide turning movement against the enemy's right flank, were met at Samarra Railway Station, which was burning. Here some fifteen engines and much rolling stock were discovered; all had been damaged, but were repairable. Two engines were got going in forty-eight hours. The 59th Rifles and 47th Sikhs put out an outpost line north of Samarra on the high banks of the Izaki Canal, which dominated the country for miles. A few Turkish patrols could be seen some miles away, retiring north, but otherwise there was no sign of the enemy, who, it was afterwards learnt, had retired some forty miles. The Regiment then marched back via Istabulat and Beled to Barura, and, having crossed the Tigris by boat bridge, became part of the reserve to the III Corps for General Marshall's operations on the Adhaim River. After marching some seventy miles the Brigade eventually returned, recrossing the Tigris at Samarra to Beled, which is only some seven miles from Barura, where the river was first crossed. Had the bridge of boats been kept a little longer, this roundabout march would have been saved. In the meantime the III Corps operations had been successful, but severe dust storms enabled the enemy to slip away once more.

Operations against Tribesmen.

On May 14th a patrol of the 47th Sikhs, escorting a British officer on reconnaissance duty from Beled, was treacherously attacked by Arabs from Samaichke, who were supposed to be friendly. The patrol was ambushed and a stubborn fight ensued, in which the British officer and seventeen or eighteen men were killed. Punitive operations against this tribe were undertaken immediately. A column consisting of the

14th Battery R.F.A., 47th Sikhs and 59th Rifles marched for Samaichke on the morning of May 17th, arriving at the village at 10 a.m. The column left Samaichke at midnight, and came up with the Arabs, who were about 1,000 strong, on the following morning. The Arabs fled as soon as the 47th advanced, but the artillery was able to inflict some casualties on them before they got away, and large flocks of cattle which they had with them were captured. The 47th Sikhs and one section of guns were left as garrison at Samaichke, while the remainder of the column returned to Beled. Further punitive operations were carried out against this tribe on May 27th, and 30 prisoners and some more cattle were captured.

The Regiment remained the whole of June at Beled, and moved to Samarra Railway Station on July 23rd, where they remained for the whole of August and September. Between June and August there were no serious operations. During this period musketry and intensive training were carried out and three large drafts arrived. In June full scale rations were obtained for the first time for some months. On September 29th news was received that Brooking's Column, operating on the Euphrates, had surrounded the Turks at Ramadi, and captured some 2,000 prisoners, including the Turkish Commander.

In September Major-General H. D. Keary, C.B., D.S.O., relinquished command of the 3rd Division, and his farewell order to the Regiment is reproduced in Appendix IV.

During October intensive training was carried on, and large fatigues were employed in building up a strong position at Istabulat, in case the enemy drove us back down the Tigris. All danger of this was past, however, after General Allenby's push at Gaza. The Brigade marched to Samarra on October 23rd, and preparations were made for fresh operations. The enemy were reported to be dug in in three lines some eight miles up the Tigris from the El Ajik position. Night operations were undertaken on the night of October 23rd-24th, but no Turks were found, although recently occupied trenches were discovered, and the Brigade returned to bivouac on October 25th. The Brigade marched at 5 p.m. on October 31st, and arrived in the El Ajik position at 9.30 p.m. The Regiment marched with the Brigade at 5.45 on November 1st. The column consisted of the entire 7th Division, the

8th Brigade, with the 4th Brigade R.F.A. and a Cavalry Division, and was operating against the enemy right flank, and therefore some distance from the Tigris. Elaborate arrangements for water supply were made, all available Ford vans being used for this. The Brigade marched until 3 a.m. on the 2nd, about three miles south-west of the Turkish position at Daur, having accomplished a night march of twenty miles. The 28th Brigade attacked the Turkish position at Daur, in the early morning, and captured it with little loss. On the 3rd the 7th Division followed up the previous day's success, and drove the enemy still farther back to his main position at Tekrit, suffering fairly heavy casualties in doing so. The Regiment marched throughout the night of November 4th-5th, and orders were received for the 8th Brigade to attack the enemy's position forthwith, supported by the 4th Brigade R.F.A. As soon as the 8th Brigade had made good its objective, the third line of enemy trenches, the 19th Brigade was to come through and exploit success.

Action of Tekrit.

The advance was begun at 6 a.m., 47th Sikhs on the left and 59th Rifles on the right, supported by the 2/124th and Manchesters. The troops came under heavy shell fire, but were advancing in artillery formation and suffered few casualties. Lieutenant G. E. Hansen was wounded almost as soon as the advance started, and Lieutenant Twinberrow took his place as commanding No. 3 Company. The Jibin Wadi was reached at 8.30 a.m., and No. 4 Company and half No. 3 Company were ordered to advance to the next nullah, some 1,500 yards ahead. Here one Company, 47th, was found, and a halt was made. The wadi where Battalion headquarters and the other two companies halted for the time being was being heavily and accurately shelled, but fortunately few casualties occurred, as the bank gave just enough cover. Orders were now received that the 47th and 59th were to attack on a bearing of 20 degrees on a frontage of 600 yards at 10.30 a.m. The 47th had to make a considerable detour to get into position, so the attack was postponed for an hour. As soon as the Commanding Officer heard this he ordered Battalion headquarters and the other two companies up to the second nullah, from which the attack was to be delivered, and there was just time for a hasty conference between the Commanding Officer and Captain Burne, commanding the firing line, before our barrage opened and the attack was launched. Nos. 3 and 4 Companies advanced out of the nullah, and after passing through heavy machine-gun and rifle fire, captured the Turkish trenches about 900 yards' distant. No. 2 Company was in support of the two front companies, and suffered heavy casualties while advancing, Lieutenant F. B. Roseveare being hit through the head before he had gone 200 yards. The company carried on under Subadar Sahib Haq, who was himself wounded later on. All three lines of trenches were captured by the 59th in twenty minutes, and the 47th had taken two lines. Nos. 3 and 4 Companies were in the third-line trench, and No. 2 in the second line, and No. 1 Company was sent into the original Turkish trench as soon as the 2/124th appeared in support of the 47th. The enemy was shelling the whole area heavily, and telephone communication was much interrupted. Information was got through to the Brigade, however, that the objectives were gained and positions firmly established. At this time (noon) Lieutenant Young, Adjutant, was severely wounded at Battalion headquarters. Captain Burne, in the front line, was also reported as wounded. The enemy made several counter-attacks, but were beaten back by Lewis-gun and rifle fire, and the 4th Brigade R.F.A. Major Anderson came up to Battalion headquarters about 3.30 p.m., and after a consultation with the Commanding Officer was sent up to the front line to send back a report on the situation.

In the meantime the 19th Brigade had not arrived. They eventually arrived at about 4.30 p.m., and occupied some trenches on the right. The Cavalry Brigade, operating on the left flank, managed to get home with a charge, but suffered severely in getting away, while mixed up in trenches and barbed wire. They were also caught in mass formation by enemy aircraft. At 4.30 p.m. the artillery opened a bombardment of the enemy trenches, and a further advance was now contemplated, but Major Anderson had not been able to complete necessary arrangements with the Manchesters by 5 p.m., and our guns had been compelled to cease fire, as the enemy had concentrated all available artillery on our gun positions. This bombardment was, as usual, the signal for an enemy retirement after dark. The enemy kept up a continual fire until 11 p.m., when all became quiet. The Turks had retired.

The Regiment was most enthusiastic over the excellent support that the 4th Brigade R.F.A. had afforded throughout the day, and the Commanding Officer was much pleased to be able to inform the artillery to this effect.

By the evening there were some 500 or 600 casualties of various units collected at 59th headquarters, and although repeated messages were sent for ambulances and water, nothing arrived until 3 a.m., and much hardship was endured by the wounded.

At 5 a.m. on the following morning the 28th Brigade passed through the 8th Brigade, which then concentrated at the Jibin Wadi and the Tigris. Captain Hamilton brought up a convoy from Samarra with blankets, etc., for the Regiment.

Casualties at Tekrit were :—

	Killed.	*Wounded.*	*Missing.*	*Slightly Wounded.*
British Officers... ...	—	4	—	—
Indian Officers ...	—	3	—	—
Indian Other Ranks ...	25	216	1	7
Total casualties			256	

The following British officers were present at Tekrit :—

Lieut.-Colonel T. L. Leeds, D.SO.
Major B. E. Anderson, D.S.O.
Major R. D. Beadle.
Captain R. H. Burne (wounded).
Captain W. H. H. Young (Adjt.), (wounded).
Lieut. J. D. Twinberrow.
Lieut. F. B. Roseveare (died of wounds).
Lieut. G. E. Hansen (wounded).
Lieut. R. S. Dudley.
Captain R. N. Kapadia, I.M.S.

At noon on November 10th news was received that Lieutenant Roseveare had died of his wounds in Samarra Hospital.

On November 15th Major Anderson and Captain Hamilton made a reconnaissance of the trenches in the El Ajik position, held by the 28th Brigade, who were relieved three days later, on the return of the 8th Brigade to El Ajik.

On the 23rd information was received of the sudden death of General Maude at Baghdad, and all British officers attended a memorial service held on the following day at the 1st Manchester Regiment's church parade.

The following immediate rewards were published in Order of the Day No. 95 on November 27th :—

Military Cross :

Acting Captain R. H. Burne.

Indian Order of Merit (*2nd Class*) *:*

Subadar Sahib Haq.

Indian Distinguished Service Medal :

3452 Havildar Devi Dyal.
4342 Havildar Niaz Gul.
2696 Naik Dhanni Ram.

During December, the Regiment remained in Camp. No active operations were carried on during the month, except for some aeroplane activity on both sides. The camp was bombed by enemy 'planes on several occasions. Great attention was paid to gas drill, as there were rumours that the enemy intended to use this in his next attack.

The following officers were serving with the Regiment at this time :—

Lieut.-Colonel T. L. Leeds, D.S.O.
Major B. E. Anderson.
Major R. D. Beadle (46th Punjabis).
Captain H. M. Hamilton.
Acting Captain E. J. Garthwaite, I.A.R.O.
Acting Captain G. E. Hansen (52nd Sikhs F.F.).
Lieutenant J. D. Twinberrow.
Lieutenant R. S. Dudley, I.A.R.O.
Lieutenant C. M. Malden.
Lieutenant D. J. Brown, I.A.R.O.
Lieutenant G. H. Simms.
Lieutenant J. L. Carter (15th Sikhs).
Lieutenant R. N. Kapadia, I.M.S.

The casualties for the year were :—

	British Officers.	*Indian Officers.*	*Other Ranks.*
Killed	2	2	74
Died of wounds ...	2	3	16
Wounded	5	5	368
Sick to F.A. ...	9	8	429

CHAPTER VIII.

THE GREAT WAR: 1918, PALESTINE.

THE Regiment saw no more active service in Mesopotamia. It was decided that the final blow at the Turks was to be delivered in Palestine, and not in Mesopotamia, so in order to reinforce the Expeditionary Force under General Allenby in Palestine, the two veteran divisions of the whole Indian Army, the 3rd Lahore and the 7th Meerut Divisions, were sent to Palestine.

On March 2nd one Indian officer and 96 other ranks of the 52nd Sikhs F.F. left to rejoin their own unit, which had now arrived in Mesopotamia. The Commanding Officer inspected this excellent detachment before it marched off, and thanked all for their gallant behaviour in every action in which they had taken part, and congratulated them on their excellent discipline during the long period they had served with the 59th Rifles F.F. These men were replaced by a draft of 83 59th men, who rejoined from the 52nd Sikhs.

The Regiment was relieved at Samarra on March 21st and in the afternoon, the 59th, 2/124th Baluchis, and Machine Gunners, were formed up on parade, and addressed separately by General Sir Stanley Cobbe, commanding the I Corps. He was much affected at saying good-bye to the Regiment, which had served continuously with the I Corps.

Departure for Palestine.

The Regiment finally arrived at Makina on April 11th, and embarked on May 9th, on the ambulance transport *Sicilia*. On the following day, men and stores were transhipped to the H.T. *Magdalena*, which arrived at Suez on May 27th, having called at Muscat and Aden. The Regiment entrained for Ismailia in the afternoon, and there joined the 3rd Division. On June 22nd, " D " Company was transferred to the 3/151st Infantry. On the following day, the 8th Brigade left for Kantara, where they entrained for Ludd, arriving there at 2 a.m. on the 26th. The 8th

Brigade marched to Wilhelma, a captured German Jew colony, and remained there as brigade in reserve. The 59th Rifles were ordered to push on to relieve the 27th Punjabis, who were acting as sector reserve to the 9th Brigade, who were in the line in front of the Wadi Belut. There was considerable artillery activity on both sides during this period. On July 18th the 8th Brigade relieved the 9th Brigade in the line, the 59th taking over the centre sub-section, Wadi Belut, from the 1/1st Gurkhas. During this relief there were some casualties, due to enemy shelling. On July 22nd, a patrol under Havildar Shamas Khan was sent out to reconnoitre a ridge about 800 yards to the left front. The patrol did not follow the route laid down for it, and was ambushed, losing two men wounded, and Havildar Shamas Khan, who was covering the retirement, was captured.

On the evening of the 22nd information was received that Lieutenants Young, Twinberrow and Kapadia had been awarded the Military Cross for operations on and about November 5th, 1917. Subadar Gauri Charan was evacuated to field ambulance on July 24th, and Lieutenant G. H. Simms took over the duties of Quartermaster.

The Regiment was relieved in the line on July 31st, and remained in camp near 3rd Divisional Headquarters until August 26th. During this period guard duties and training were carried out. They went into the line again, on August 26th, relieving the 2/124th at Rasolain, an old Crusader castle, very celebrated, and a most prominent landmark. On September 6th a patrol of " B " Company, under Lieutenant Carter, which went up the metre gauge railway, to reconnoitre the enemy's line, was ambushed by the enemy, and reported five men missing on return to Rasolain.

On September 15th all Commanding Officers of the 7th and 8th Brigades attended at Divisional Headquarters to meet General Allenby, who made a very brief speech thanking all Commanding Officers for the trouble they had taken in making preparations for the forthcoming operations.

Final Advance in Palestine.

He said he was sure that all Commanding Officers would take even more trouble over the actual operations, and that the 3rd Division would break through the Turkish line and be in Nablus on the third day after the break through. The cavalry would go through the gap made by the infantry, and would capture the entire Turkish forces in Palestine.

The foregoing is exactly what actually happened.

On September 17th preliminary orders for ensuing operations were issued to Company Commanders. On the following day Brigadier-General Edwardes visited the Regiment and made a speech to all officers, in which he said that he felt quite confident that things would go well, and that the Regiment would do well in the coming push. He wished everyone the best of luck. At 6.30 p.m. the Regiment moved to its allotted station, where it remained the night in readiness for next day's operations.

The plan of operations was briefly for the 3rd and 7th Divisions to assault and break through the Turkish lines immediately opposite them, the 7th Division being on the left of the 3rd. These two divisions held the line from the metre-gauge railway westwards towards the sea. As soon as the gap was made, both divisions were to change front right and push on as hard and as fast as possible due east. The cavalry was to follow these two divisions through the gap they had made, and proceed due north as fast as possible for about fifteen miles, and then change front right, cutting off the enemy's line of retreat. Zero was 4.30 a.m. September 19th, at which time an intensive bombardment by our massed guns opened on the line to be assaulted. The assaulting troops had rehearsed this assault on similar ground in rear, and knew their job thoroughly. The 8th Brigade was in Divisional reserve for the actual assault, but the Manchesters and the 2/124th had minor rôles assigned to them, the Manchesters to attack and capture Jiljulia, and the 2/124th to capture an enemy position immediately in front. The Manchesters' job was stiffer than was anticipated, but the 2/124th succeeded at once, and later took Railway Redoubt, from which they were able to support the Manchester attack. At 10.45 a.m. the 59th were ordered forward to Jiljulia and Hableh. The Regiment advanced in artillery formation and reached Hableh at 1 p.m., and changing front right, passed through the Manchesters, to attack Rasel Tireh. At 4 p.m. the Regiment advanced in diamond formation, "B" and "C" Companies leading. At 5.20 p.m. "C" Company reported the capture of two Turkish guns, and "B" Company had captured Rasel Tireh with very few casualties ; the enemy had been driven back all along the line. It was now dark, and all ranks were exhausted, so that it was decided to put out outposts for the night, and to continue the advance next morning. The 54th Division were on the right flank, but had not come up far enough for touch to be established. The 2/124th were on the left.

At 4 a.m. on September 20th, orders were received that the Manchesters were to pass through the 59th, who were to follow in support. Meanwhile the 47th Sikhs had passed through the 2/124th and advanced among the hills on the left across the Wadi Azun. The 1st Manchesters drove the enemy out of their position after some opposition, and occupied Kefir Tilth. The 59th then pushed on and cleared the enemy off their next position in the hills after considerable opposition. At noon the advance was held up by a long ridge across a wadi on the left flank, which was held by the enemy, and enfiladed our position. At 12.30 the batteries had registered, and " A " and " C " Companies were ordered to move round and attack the enemy's left flank up the spur. A few rounds from the guns, and the threat to his line of retreat caused the enemy to evacuate his position, after delivering his customary long bursts of rapid and machine-gun fire.

The Regiment now found itself in very broken and hilly country, and all were thoroughly tired, but the orders were to pursue the Turkish rearguard as fast as possible, and the advance was therefore continued. Touch was established with the 47th Sikhs in the afternoon, and the enemy were found in position between two hills, which were connected by a col over which the road ran. The enemy were pushed out of this position, and that night both the 7th and 8th Brigades bivouacked at Jinsafat, about a mile beyond the two hills. During the day the Regiment lost 2 men killed and 16 wounded. On the morning of the 21st, the advance was continued, and many abandoned guns and ordnance and transport wagons were passed on the road. The troops came under long-range shell fire at about three miles from Nablus, and at this point the 8th Brigade was ordered to Samaria, across the Grand Trunk Road of Palestine, running north from Jerusalem. This road was the main line of the enemy's retreat. Here a squadron of French cavalry was met. They had just been in contact with the flying enemy and a great quantity of abandoned carts and guns were seen on the road.

The Brigade arrived at Samaria in the afternoon, and bivouacked there the night, moving south to Zawata on the following day. News was received that up to date 130 guns and 8,000 prisoners had been captured. On the evening of the 22nd reinforcements and 20-lb. kits arrived, and everyone was glad to get a blanket, as the nights had now turned very cold. The XXI Corps saw no more fighting after this. The cavalry and Royal Flying Corps were in contact with the flying enemy, who were all north of Nazareth by this time.

The following officers were present during the final advance :—

Lieut.-Colonel T. L. Leeds, D.S.O.	A./Captain R. S. Dudley.
Major B. E. Anderson, D.S.O.	Lieut. C. M. Malden.
A./Captain G. E. Hansen.	Lieut. J. L. Carter.
A./Captain J. D. Twinberrow, M.C.	Lieut. A. F. Telfer.

The Regiment arrived back at Jiljulia on September 29th, and was employed chiefly on salvage work. On October 31st news was received that an armistice had been signed with Turkey, and news of the armistice with Germany was received on November 11th.

The Regiment remained in camp at Ludd without incident from December, 1918, until March 2nd, 1919, when the advance party for India under Major Anderson and Lieutenant Twinberrow left for Suez, under orders to embark for India. On March 21st General A. R. Hoskins commanding the 3rd Division, inspected the Regiment at Ludd station, previous to entraining for embarkation to India. He made an exceedingly flattering speech, and was heartily cheered by the Regiment as they left the station.

Egyptian disturbances.

The Regiment arrived in camp at Suez on March 22nd, but the Egyptian disturbances had now broken out, and all units in the transit camp were immediately made responsible for the defence of Suez. The Regiment took over all Suez guards and piquets on April 10th, as the 47th Sikhs and 58th Rifles were under orders to proceed by train, destination not stated. Battalion headquarters were moved from the camp to the civil hospital in Suez, and garrisons were supplied at Port Tewfik, Suez, and the oil refinery, with guards on all important places in the area. There were some disturbances and riots at Suez, and some firing by the police occurred, but it was never found necessary by the Regiment to resort to firing, all crowds dispersed whenever the troops turned out. Agitators from Cairo came to Suez, and tried to stir up trouble, but the local people evidently considered that the situation was too well watched for them to be able to do anything. Some trouble occurred among the coolies at the oil refinery, but the excitement was soon stopped, and all became quiet.

The Regiment remained in Suez until May 7th, when it was moved to Ismailia, and was put on duty in connection with the French canal strike. At the end of May orders were received for the 59th to rejoin the 3rd Division in Palestine. It was naturally a bitter disappointment to all ranks to learn that the Regiment which had now been almost five years overseas, and which had been a firing-line regiment all the time, was now to return to its old division, after having expected to go back to India. However, all ranks accepted the fact, and behaved in a most satisfactory manner over their disappointment. G.H.Q. at Cairo at once granted the Commanding Officer's request for leave to India to be opened, and a party was sent off immediately.

The Regiment entrained at Kantara on June 8th, and went into camp at Bir Salem on the following day, and remained here until December 8th, when it arrived at Jerusalem. "A" and "B" Companies proceeded to Goronya in the Jordan Valley, and "C" Company marched to Hebron. Lieut.-Colonel T. L. Leeds took over command of the Ludd Area, and Captain H. M. Hamilton took over command of the Regiment.

On July 1st orders were received that each Indian battalion in the Egyptian Expeditionary Force should send one Indian officer, one non-commissioned officer, and one sepoy to take part in the peace celebrations in London, and Subadar Sahib Haq, I.O.M., No. 1421 Havildar Rijha, and No. 395 Sepoy Abdul Khanam were selected to represent the Regiment.

The 8th Jullundur Brigade, 1912-1919.

It is impossible to close the story of the 59th Rifles in the Great War without reference to the continuous association throughout the war of the 1st Battalion The Manchester Regiment, the 47th Sikhs, and the 59th Scinde Rifles, F.F., who served together in the 8th

Jullundur Brigade from 1912 to 1919. The great friendship and liaison which existed between these three battalions contributed largely to the magnificent reputation gained by the Brigade.

To commemorate such splendid association, three identical centre-pieces were ordered from the Goldsmiths and Silversmiths Company, the centre piece consisting of a triangular column rising from a triangular base, surmounted by a winged figure of Victory. At each corner of the base there is a silver model of a soldier of each battalion. Each battalion is now in possession of one of these centre-pieces, presented by the remaining two battalions.

59th Memorial to Officers who fell in the Great War, erected in Kohat Church.

Centrepiece to commemorate the continuous service throughout the War of the 1st Manchesters, 47th Sikhs and 59th Rifles in the Jullundur Brigade.
Each battalion has one piece, presented by the remaining two battalions.

CHAPTER IX

1920–1923

At the beginning of 1920 the Regiment was still at Jerusalem. During January " A," " B " and " C " Companies rejoined from detachment duties, and the Regiment left Jerusalem for Kantara, arriving there on February 1st. Lieutenant-Colonel T. L. Leeds, having been ordered to proceed with the Regiment to India, relinquished command of the 8th Brigade at Ludd, and rejoined the Regiment. The Indian Transit Camp at Suez, where the Regiment arrived the year before, was reached on March 20th. Here the Regiment remained until May 5th, when it embarked on H.T. *Swakepmund*, a captured German ship, sailing the same day. Aden was reached on May 12th, and Bombay on the 19th.

The Regiment arrived at Jullundur at 9.30 p.m. on May 23rd. On the following day the Battalion was inspected by the Commissioner for the Jullundur District, who addressed the Regiment. All men were then sent on two months' war leave as soon as pay could be issued.

On May 25th the Regiment and the Depot were amalgamated. The following officers were present :—

Lieut.-Colonel T. L. Leeds, C.M.G., D.S.O.
Captain C. E. Stuart-Prince.
Captain G. E. Hansen, Adjutant.
Captain J. D. Twinberrow, M.C.
Lieutenant C. M. Malden.
Lieutenant D. M. Lindsay.
Lieutenant J. W. Banks.
Lieutenant R. E. Wilson.
Lieutenant T. F. S. Sawyer.
2/Lieutenant F. Ashcroft.

Captain H. M. Hamilton and Lieutenant G. H. Simms, Quartermaster, rejoined five days later.

On August 21st the Regiment entrained for Kohat, arriving there on the following day. They relieved the 9th Bhopals in the Lockhart Lines.

On November 14th a daring raid was carried out in Kohat, under Mirz Ali, a notorious Afridi outlaw, resulting in the murder of Colonel and Mrs. Foulkes. The raiders got away scot free at the time, but it was hoped to capture some of them later, as the members of the gang were all known. One was caught about a year later and brought to justice.

During the year the undermentioned well-known old Indian officers were pensioned :—

*Honorary Captain Nasir Khan, Sirdar Bahadur, I.O.M. ;

Honorary Lieutenant Makhmad Jan, Sirdar Bahadur, I.D.S.M.

These were fine types of the old pattern, long service Indian officer, whose presence was invaluable, and who could with difficulty be replaced. Both had served in the Regiment for thirty-seven years, and had done invaluable work at the Depot, after having returned from France in 1915.

A great number of men were demobilized during the year, and at the end of 1920 the Regiment consisted chiefly of young sepoys and inexperienced non-commissioned officers, not yet fully trained. The Indian officers were also mostly young.

Lieut.-Colonel T. L. Leeds, C.M.G., D.S.O., handed over command of the Regiment to Major C. R. Wilkinson, D.S.O., on February 8th, 1921, on which date he left Kohat on leave *ex* India, pending retirement. Lieut.-Colonel Leeds had commanded the Regiment since February, 1915, and had been with it as Commanding Officer during all the heavy fighting in France, Mesopotamia and Palestine. A farewell party was given to Colonel Leeds on the afternoon of February 7th, during which the final of the Company football tournament was played off for the Leeds Shield, which was won by " A " Company.

On February 11th news was received that the Regiment had gained the distinction of the title " Royal." Congratulatory telegrams began

* The former of these two old Indian officers died at his home at Swabi in 1923. His son, Jemadar Shamas Khan, was granted a direct commission in the Regiment in 1919. Makhmad Jan was made a Khan Sahib by the civil authorities in 1922, after he had rendered fine services in the Kohat levies which were raised during the disturbances of 1919-20. Since retiring he has rendered the authorities invaluable help in operations against raiders.

to pour in, and letters or telegrams were received from His Excellency the Commander-in-Chief, Lord Rawlinson, Sir Charles Monro, late Commander-in-Chief in India, General Birdwood, and General Edwardes. Colonel Leeds received the news just before he embarked at Bombay.

Reorganization.

During February, 1921, orders were received to change the composition of the Regiment. These orders affected every unit in the Indian Army. Regiments were grouped together, with a Training Battalion for each group, the 59th being grouped with the 55th, 56th, 57th and 58th, with the 2/56th Rifles as the Training Battalion. The group number was 13, the Regiment's official designation being 59th Royal Scinde Rifles (Frontier Force), 5th Battalion, 13th Indian Infantry Group.

The change of composition resulted in the Regiment losing a half company of Pathans (Khattaks) and taking a half company of P.Ms. in their place. It was fortunately possible to find a vacancy for the whole of the two platoons of Khattaks with the 52nd Sikhs F.F., and they were transferred there during March. The loss of the Khattaks, under Subadar Khan Gul, I.O.M., was greatly regretted by everyone in the Regiment.

Orders were also received about this time to take on a platoon of Afridis, to make place for whom a proportionate amount of men in the Pathan Company would have to be discharged. The 59th were ordered to take Kambar Khel Afridis, and towards the end of August the first batch of recruits were obtained, a good proportion of whom, in spite of being vouched for by the Political authorities, turned out to be bad characters, deserters, etc., and had to be discharged. The composition of the Pathan Company became one platoon Kambar Khel Afridis, one platoon Yusafzais, one platoon Upper Bangashes, and one platoon Lower Bangashes.

Major C. R. Wilkinson, D.S.O., was promoted Lieutenant-Colonel on February 1st, and was appointed Commandant of the 59th, with effect from August 20th, 1921.

During the hot weather, 1921, a Headquarter Company was formed in accordance with the new organization, and the strength of the Battalion was fixed at 20 Indian officers and 806 Indian other ranks. The Vickers gun platoon, No. 20, was made up of all classes, one team for each gun being found by each class company. New Regimental Standing Orders were made out and issued.

Operations against Outlaws.

On November 25th the Regiment furnished a party of 200 rifles from "A" and "B" Companies, with a headquarter platoon from "D" Company, under Captains Duncan and Wainwright, which co-operated with the Frontier Constabulary from Hangu, and succeeded in capturing nine outlaws. The operations involved motoring in lorries to Doaba, and then a night march to Dumbaki, a distance of eleven miles. The outlaws were captured in this village, which was surrounded before dawn.

On November 30th "A" and "B" Companies proceeded to Dhoda for company training, returning on the 21st. This was the first company training under the old system carried out since 1913. "C" and "D" Companies and the Vickers Gun Platoon went out to camp in January.

The formation of the two new platoons of P.Ms., in place of the Khattaks transferred, was practically completed by the end of the year, but as recruiting was not open, transfers from units disbanding, etc., had to be taken. The result was not on the whole satisfactory.

On March 11th, 1922, the Regiment proceeded to Hangu and the Samana, taking over the posts of Fort Lockhart, Gulistan and Sangar from the 29th Punjabis.

On May 25th news was received of the death of Captain J. S. Culverwell in Simla, where he had taken up a staff appointment in the previous year. The Regiment suffered a severe loss in this officer, who was above the average in capacity, and had seen a lot of service with the 59th in the war 1914-19, being twice wounded, and having received promotion for gallantry in action.

At the beginning of November orders were received that the Regiment was to proceed to the Persian Gulf, after being relieved on the Samana by the 122nd Rajputana Rifles. The sending of all ranks on overseas leave was begun, when orders were received to suspend operations until further instructions. In the meantime the relief was held up, and the 122nd were established on the Samana, and had a Company at Hangu, which had not yet, however, been handed over to them. On December 19th orders were received to carry on with the furlough and relief, and information was received that the Regiment would proceed if necessary to relieve the 15th Sikhs in Iraq. Definite orders would be issued during January. In the meantime a cadre was kept in Hangu. The training of the Regiment was seriously interfered with during the year by the abnormal amount of leave granted, practically every man receiving five and a half months' leave between March, 1922, and April, 1923, and also by the fact that the Regiment was split up into detachments.

On December 2nd the Regiment received the new title of 6th Royal Battalion 13th Frontier Force Rifles (Scinde). The other Regiments in the 13th Group received Battalion numbers beginning with the 55th Rifles as the 1st Battalion, the 3rd being left out, so that all battalions gained their old Punjab Infantry numbers.

The last batch of recruits trained under the old system with the Regiment was passed out in the hot weather of 1922. From this date all recruits were enlisted and drilled at the Training Battalion.

The following officers, many of whom had been with the 59th for some years during the war, retired under the Royal Warrant during 1922 :—

Captain J. D. Twinberrow, M.C., 1915-20.
Captain G. H. Simms, 1916-22.
Captain A. J. Hobley, 1917-18, 1921-22.
Captain H. N. Lindsay, 1918-22.
Lieutenant G. D. Florence, 1918, 1921-22.
Lieutenant W. R. Darnell, 1921.

Lieutenant T. F. S. Sawyer (1919-23) retired at the beginning of 1923.

Captain C. E. Stuart-Prince, who had joined the Regiment at the beginning of 1914, retired under the Royal Warrant in April, 1923. For over two years previous to his retirement he had commanded "B" Company.

Daring Raid on Kohat.

In March, 1923, orders were received that the Regiment would proceed to Iraq. Just before the Regiment was to leave Hangu, Kohat was raided and Mrs. Ellis was killed, while her daughter was carried off by the raiders. The 59th received orders to stand by early on the following morning, and at 12.30 p.m. moved out from Hangu with no rations, transport, or Lewis or machine guns. The men had time to cook a hasty meal before moving off.

The Battalion moved up the Bar Valley towards Kohat, covering a tract of country some three miles wide. A line of piquets was put out that night midway between Hangu and Kohat. The line was stretched over very broken country, and inter-communication was extremely difficult, all signalling stores having been returned to Ordnance, and the best arrangements possible having to be made with a mixed collection of equipment hastily borrowed from the 56th Rifles, who had just arrived in Hangu as relief. Hangu was also out of touch.

After a cold night, spent on the hillside, the Regiment advanced on the following morning, and gained touch with the Kohat Brigade. There was no further information as to the whereabouts of the raiders, and the troops returned to Kohat and Hangu that afternoon. All the Mohammedans of the Regiment were most kindly provided with food and tea by Subadar-Major Nauroz, formerly of the Kurram Militia, in the village of Ibrahimzai, on the return march to Hangu.

Troops continued operations from Kohat, but no trace of the raiders was discovered. Some days later Miss Ellis was safely brought back to Kohat, mainly through the efforts of Mrs. Starr, of Peshawar, who proceeded across the border to where Miss Ellis was held in captivity. As it turned out, this daring raid was merely the prelude to numerous further outrages that took place on the frontier in 1923.

During the winter 1922-23, Major B. E. Anderson, D.S.O., served with the 56th Rifles (Frontier Force), and Major A. H. Burn, C.I.E., O.B.E., served as D.A.A.G. with the Razmak Field Force, both officers returning to proceed with the Regiment to Iraq. Both officers were subsequently mentioned in despatches for their work with the Razmak Field Force.

Lieut.-Colonel A. E. Mahon, D.S.O., joined the Regiment as Second-in-Command in February, 1922, but was transferred to the 1/12th Frontier Force Regiment (Sikhs) as Commandant in the following year.

Major R. D. Inskip, D.S.O., M.C., rejoined the Regiment in February, 1923, at Hangu, having graduated at the Staff College, Camberley.

The Battalion left Hangu for Kohat on April 18th and embarked at Karachi on the B.I. boat *Varsova* on the 22nd, after having served in the 6th Indian Infantry Brigade at Kohat since August, 1920.

Arrival in Iraq.

On arrival in Iraq the Battalion came under the orders of the Royal Air Force, and information was received that we should be stationed at Basra. Almost immediately, however, two companies were ordered to proceed to Baghdad temporarily, to replace troops despatched against Sheikh Mahmud. "B" and "C" Companies, under Major A. H. Burn, left for Baghdad on April 28th.

Four days later orders were received for the remainder of the Battalion to proceed to Baghdad. This move was most welcome, on account of the escape from the Basra hot weather, and in view of the fact that the Battalion at Basra at that time provided large detachments at Shaibah and Nasiriyeh, while the duties in Basra were extremely heavy.

At about this time news was received that General Sir James Willcocks, G.C.B., G.C.M.G., K.C.S.I., D.S.O., had been appointed Colonel of the Regiment. This was regarded as a great honour by all ranks, especially by those who had known and served under General Willcocks when he commanded the Indian Corps in France. In September the following letter was received by the Commanding Officer from Sir James Willcocks :—

"This is only a short note to tell you and all ranks of your distinguished Corps how pleased and proud I am at being made Colonel of the Regiment. I have served so often both in peace and on service with nearly every Battalion of the grand old Punjab Frontier Force, that I deem it a special compliment that you should have done the honour of asking for me. When I know who is in command now and the proper station of the Corps I will write again. I wish it to be understood that I do not desire to be your Colonel in name only, but I shall be in a position sometimes to help in matters connected with the War

Office or Army Headquarters in India, and if at any time I can do this, I am anxious and willing to do so, as I do now for the British Regiment of which I am Colonel.

"Every good wish for all ranks, and please convey to the Indian officers and men my earnest hope that I shall see the Battalion on its return to India."

On arrival at Baghdad the Battalion was quartered in extremely comfortable lines in Baghdad West, with three platoons in Baghdad supplying the headquarter guards there. The 1st Battalion Frontier Force Rifles (late 55th) were at Hinaidi, and left for duty with the column two days after our arrival.

The Regiment remained in Baghdad West until August, 1923 during which time individual training was carried out without interruption. Nearly every man was taken up for a short period in troop carrying aeroplanes, which had recently been successfully used in operations. The aeroplanes were flown over from Hinaidi especially for the purpose of giving men these flights.

During May news was received that Major B. E. Anderson, D.S.O., had been promoted Brevet Lieut.-Colonel.

In August the 1st Battalion Frontier Force Rifles left for India, and the Regiment marched to Hinaidi and took over the Indian infantry lines there. These lines were newly built, and were well fitted out. Our duties in Baghdad West and later in Baghdad itself were taken over by the Iraq Levies.

During the cold weather 1923-24, two company training camps were held, one at Lajj, and the other at Qutniyeh. At each of these camps, No. 6 Squadron, Royal Air Force, provided a liaison Flight, with which much valuable and instructive co-operation work was done. Some Indian Officers and nearly all the signallers were taken up for flights in D.H.9 machines, and gained some experience of methods of communication between aircraft and infantry. Later in the cold weather Lieut.-Colonel B. E. Anderson, D.S.O., conducted a tour for British officers at Table Mountain. A few officers of the Royal Inniskilling Fusiliers also attended this tour.

A platoon hockey tournament was held during June, July, and August, and was won by Subadar Bhan Singh's Platoon.

LAST BATCH OF RECRUITS TRAINED UNDER THE OLD SYSTEM WITH THE REGIMENT.

Taken at Hangu in March, 1922. Capt. D. M. Lindsay, S.M. Perbhat Chand, M.C., J.Adjt. Niaz Gul, I.D.S.M.

These recruits were mostly Afridis and Sikhs, the former being newly-enlisted men to make up the Afridi platoon which had just been ordered to be raised, while the latter were chiefly demobilized men of other units re-enlisted to make up the Sikh company strength. All these recruits had passed off the square by July, 1922.

BRITISH AND INDIAN OFFICERS AT HINAIDI, BAGHDAD, 1924.

Lieut.-Colonel C. R. Wilkinson, D.S.O., Commanding. Subadar-Major Perbhat Chand, M.C., on Col. Wilkinson's right.

In October and November companies proceeded to a training camp at Ctesiphon, two at a time for a period of three weeks.

Return to India. The Battalion moved to Basra in January, 1925, and was relieved in March by the 2/16th Punjab Regiment, and eventually embarked for India on the s.s. *Varela* on March 15th.

Lieut.-Colonel C. R. Wilkinson, D.S.O., after having completed four years in command of the Regiment, was placed on the unemployed list on January 31st, on which date he handed over command to Lieut.-Colonel B. E. Anderson, D.S.O. Colonel Wilkinson had arrived to take over command at a period when the Battalion had only recently arrived in Kohat after return from six years' service overseas. Demobilization was still going on, the Battalion was surplus to establishment, including five Indian officers, and the reorganization scheme, involving a change in the composition, was about to be introduced. These difficulties were all eventually overcome, and reorganization was complete before the Regiment left Hangu. The personnel of the Indian hospital assisted the Indian officers in providing a farewell entertainment by staging an amusing play in the M.T. theatre.

All ranks of the Battalion will always have pleasant recollections of their service under the Royal Air Force. We had always been extremely well looked after, and the men had been made most comfortable. Quarters, rations, and lighting were better than anything obtainable in India, and R.A.F. transport arrangements were always excellent. Extra overseas pay was granted, and all men had substantial balances standing to their credit, on leaving the country. It might be added that the Royal Air Force pay authorities were always most helpful, and arrangements in this respect were excellent.

The Regiment had been most fortunate in spending almost the whole of its tour of duty in Baghdad, where it had remained together the whole time, and where much experience was gained of the work of the Royal Air Force.

On the departure of the Battalion from Iraq, the Air Vice-Marshal flew down specially from Baghdad to say good-bye, and sent a farewell telegram on the day the Battalion sailed from Basra.

In April, 1924, Subadar-Major Perbhat Chand, M.C., went on pension, having served as Subadar-Major since 1915. He had been with the Regiment throughout the war 1914-18, and had a most distinguished record. At the Battle of Neuve Chapelle he had commanded the Regiment in action against the Germans, after all British officers had become casualties. He was succeeded as Subadar-Major by Subadar Pertab Singh, nephew of the late Subadar-Major Ditt Singh.

The Regiment arrived at Delhi New Cantonments on March 23rd, and immediate preparations were made for sending all the men on three and a half months' furlough.

APPENDICES

K 2

APPENDIX I

COPY OF A LETTER FROM HIS EXCELLENCY MAJOR-GENERAL SIR CHARLES NAPIER, K.C.B., GOVERNOR OF SCINDE.

To:

THE RIGHT HONOURABLE LORD ELLENBOROUGH,
&c., &c., &c.

Dated HYDERABAD,
June 20th, 1843.

MY LORD,

I have received a letter from Captain Thomas from Bhawalpore by which I perceive that your Lordship has written to me on the subject of a Camel Corps for the purpose of carrying Infantry. I have never received this letter.

I have not the least doubt of the utility of such a Corps, in this country especially. In fact I made one of the 22nd Regiment when we marched to Emani Ghar. Each camel carried two soldiers, who walked and rode alternately. Occasionally both rode, but this was inconvenient.

Had I such a Corps with me a few days ago, I dare say we would have captured Sher Mahomed. It gives Infantry the rapidity of Cavalry, and they are enabled to go into action unfatigued after making a march of forty miles. Should a guerrilla warfare arise in this country, I have always had the intention of attacking the enemy in this manner by small moveable columns.

The nullahs are always difficult for camels to cross, but it must be recollected that it is the loaded camel that incurs so much danger. The animal that has only his own weight to carry can much more easily pass them, and if he falls does not dislocate his hip joints.

I have, &c.,

(Signed) C. I. NAPIER,
Governor.

APPENDIX II

List of Commandants of the Regiment

Lieutenant R. FitzGerald	1843-1849
Major G. B. Michell	1849-1854
Lieutenant C. P. Keyes	1854-1858
Captain W. D. Hoste	1861-1873
Major B. R. Chambers	1873-1883
Lieut.-Colonel S. J. Browne	1883-1890
Lieut.-Colonel A. N. Sandilands	1890-1892
Major J. E. Mein	1892-1901
Lieut.-Colonel E. W. Cunliffe	1901-1908
Lieut.-Colonel R. A. Carruthers	1908-1912
Lieut.-Colonel P. B. B. Forster	2/3/13-25/8/13
Lieut.-Colonel C. C. Fenner	25/8/13-1914
Lieut.-Colonel P. C. Eliott-Lockhart	1914-1915
Lieut.-Colonel T. L. Leeds	1915-1921
Lieut.-Colonel C. R. Wilkinson	1921-1925
Lieut.-Colonel B. E. Anderson	1925

APPENDIX III

Succession Roll of British Officers

NAME, WITH RANK ON JOINING.	DATE OF APPOINTMENT OR JOINING.	BECAME NON-EFFECTIVE.	CAUSE.	REMARKS.
Lieut. R. FitzGerald	1843	1849	To 5th Punjab Cavalry	1st C.O., 1843-1849.
Lieut. F. F. Bruce	1843	—	—	1st Adjutant.
Lieut. H. Bruce	1843	1849	To 5th Punjab Cavalry	—
Major G. B. Michell	1849	1854	—	C.O., 1849-54.
Ensign W. H. Paget	1849	1852	To 5th Punjab Cavalry	—
Ensign A. W. Graham	1850	1852	—	—
Lieut. J. Moore	1851	1856	Invalided	Adjutant, 1851-54.
Lieut. C. P. Keyes	1854	1858	To 30th Native Infantry	C.O., 1854-58.
Lieut. T. Quin	1854	1868	—	Adjutant, 1854-57.
Lieut. G. N. Saunders	1857	1861	To 3rd Punjab Infantry	Adjutant, 1858-61.
Capt. W. D. Hoste	1861	1872	Died at Kohat, 2/10/1872	C.O., 1861-72.
Lieut. S. J. Browne	1861	1890	Retired	Adjutant, 1861-64 ; C.O., 1883-90.
Lieut. W. C. Chowne	1864	1885	To 2nd Punjab Infantry	Adjutant, 1864-68.
Lieut. T. F. Bruce	1864	1887	Retired	Adjutant, 1868-78.
Lieut. J. F. P. Mosely	1865	1869	To 3rd Punjab Infantry	—
Lieut. C. B. Norman	1868	1869	To 1st Sikhs	—
Lieut. A. N. Sandilands	1869	1892	Retired	C.O., 1890-92.
Ensign A. G. Yaldwyn	1869	1876	To Commissariat	—
Ensign F. D. Battye	1869	1871	To Corps of Guides	—
Lieut. C. C. Egerton	1871	1873	To 4th Punjab Cavalry	—
Major B. R. Chambers	1873	1883	On vacating command	C.O., 1873-83 ; Colonel of the Regiment.
Lieut. W. H. Sim	1873	1875	—	—
Lieut. H. B. Urmston	1875	1888	Killed in action, Black Mountain, June 18th	Adjutant, 1878-82
Lieut. E. B. J. Vaughan	1877	1887	Killed Burmah, Feb. 14th	—
Lieut. E. W. Cunliffe	1882	1908	Retired	Adjutant, 1882-85 ; C.O., 1901-08.
Lieut. G. R. MacMullen	1883	1902	—	—
Lieut. D. J. O. Taylor	1885	1906	Retired	Adjutant, 1885-86.
Lieut. J. W. C. Hutchinson	—	—	—	Adjutant, 1886.
Lieut. C. Chamier	1887	1891	To Political Department	—
Lieut. K. D. Erskine	1887	1894	To Political Department	—
Capt. J. E. Mein	1887	1901	Retired	C.O., 1892-1901
Lieut. E. H. Bemard	1887	1898	To Cantonment Magistrate's Department	—

APPENDIX III—Succession Roll of British Officers—*continued.*

Name, with rank on joining.	Date of appointment or joining.	Became non-effective.	Cause.	Remarks.
Lieut. W. S. Browne	1888	1893	Died of enteric fever, Camp Malana, July 20th	—
Lieut. C. C. A. Sillery	1890	1898	To Burmah Military Police	Adjutant, 1890
Lieut. F. D. Grant	1890	1897	To Military Accounts Dept.	—
Lieut. A. Limond	1892	1895	Killed, Waziristan, May 13th	Adjutant, 1894-95.
Lieut. C. C. Fenner	1893	1914	Killed in action, Dec. 23rd	C.O., 1913-14.
Lieut. T. C. Plowden	1893	1894	To civil employ	Adjutant, 1893-94.
Lieut. A. E. Dallas	1893	1894	To S. & T. Corps	—
Lieut. F. D. Browne	1894	1905	To 56th Rifles, F.F.	Adjutant, 1899-1903.
Lieut. E. Kirkpatrick	1895	1920	Retired	Adjutant, 1895-99.
Capt. A. Nicholls	1896	1896	—	—
Lieut. T. L. Leeds	1896	1921	Retired	C.O., 1915-21.
Lieut. A. G. Ames	1896	1896	To 2nd Sikhs	—
Lieut. H. de C. O'Grady	1898	1919	To 2nd Sikhs	—
Lieut. T. J. Willans	1899	1910	To 57th Rifles, F.F.	—
2/Lieut. R. N. S. Gordon	1899	1900	To 2nd Punjab Infantry	—
Lieut. J. Clementi	1900	1900	To Q.V.O. Corps of Guides	—
Lieut. T. McC. Nicholson	1900	1906	To S. & T. Corps	Adjutant, 1903-06.
Lieut. H. W. Martin	1901	1918	Died at Jullundur	—
Major R. W. Falcon	1901	1901	To 4th Sikhs	—
Capt. W. N. Campbell	1901	1903	Retired	—
Lieut. A. H. Arbuthnot	1902	1908	To 2nd Gurkha Rifles	—
Capt. P. B. B. Forster	1902	1908	To 52nd Sikhs, F.F.	Returned as C.O. 2/8/13-25/8/13.
Lieut. A. M. Gillies	1902	1906	Died as result of accident at Polo, Sept. 30th	—
2/Lieut. W. Campbell	1902	1905	To S. & T. Corps	—
Lieut. K. D. B. Murray	1903	1921	To 55th Rifles, F.F.	Adjutant, 1906-10.
Lieut. B. E. Anderson	1904	—	—	Adjutant, 1910-14; C.O., 1925.
Major C. G. Prendergast	1904	1904	To 57th Rifles, F.F.	—
Major F. J. H. Wynch	1906	—	—	—
Lieut. C. T. C. Plowden	1906	1908	To civil employ	—
Lieut. H. F. D. Stirling	1906	1917	Killed in action, Jan. 9th	—
Lieut. H. N. Lee	1906	1914	Killed in action, Dec. 19th	—
Lieut. A. H. Burn	1906	—	—	—
2/Lieut. R. D. Inskip	1906	—	—	Adjutant, 1914-16.
Major R. A. Carruthers	1907	1913	To staff employ	C.O., 1908-12.
Lieut. W. F. Scott	1908	1914	Killed in action, Oct. 25th	—
Lieut. H. N. Urmston	1908	1914	Wound pension	—

2/Lieut. J. C. Atkinson	1909	1914	Killed in action, Dec. 19th	—
V.C., 2/Lieut. W. A. McC. Bruce	1911	1914	Killed in action, Dec. 19th	—
2/Lieut. J. A. M. Scobie	1911	1916	Killed in action, March 8th	—
2/Lieut. C. E. Stuart-Prince	1914	1923	Retired under Royal Warrant	—
2/Lieut. H. M. Hamilton	1914	—	—	—
Capt. E. A. Trafford, 52nd Sikhs F.F. ...	10/11/14	6/9/15	—	—
Capt. C. R. Gilchrist, 46th Punjabis ...	21/11/14	19/12/14	Killed in action 19/12/14	—
Capt. J. D. Scale, 84th Punjabis	21/11/14	19/12/14	Wounded	—
Capt. P. S. Hore, 52nd Sikhs F.F.	8/12/14	12/3/15	Killed in action 12/3/15 ...	—
Capt. S. R. Shirley, 54th Sikhs, F.F. ...	8/12/14	12/12/14	—	—
Capt. E. C. Barnes, 19th Punjabis	31/12/14	11/3/15	Wounded	—
Capt. H. C. Fielding, 38th Dogras	31/12/14	11/7/15	Killed in action, 11/7/15 ...	—
Capt. B. Clerk, 82nd Punjabis	5/1/15	12/3/15	Killed in action, 12/3/15 ...	—
Capt. T. Reed, 67th Punjabis	5/1/15	12/3/15	Killed in action, 12/3/15 ...	—
Lieut. M. H. Bickford, 38th Dogras ...	8/1/15	30/3/16	To duty on Lines of Communication.	
Lieut.-Colonel P. C. Eliott-Lockhart ...	29/1/15	12/3/15	Killed in action, 12/3/15 ...	—
Capt. J. R. Wynter, 52nd Sikhs, F.F. ...	9/2/15	8/3/16	Wounded	—
Capt. L. W. Passy, 25th Punjabis	1915	—	Temporary duty after Battle of Neuve Chapelle, 16/3/15 to 30/3/15	—
Capt. G. W. Atkins, 25th Punjabis ...	1915	—		
Capt. N. C. Sparling, 54th Sikhs, F.F. ...	1915	—		
Capt. W. B. Hore, 120th Infantry	1915	—		
Capt. E. W. H. Marsh, 13th Rajputs ...	1915	—		
2/Lieut. J. A. Shelverton, I.A.R.O. ...	20/3/15	26/4/15	Wounded	—
Capt. H. G. Turner, 106th Pioneers ...	21/3/15	1/4/16	To 106th Pioneers ...	—
Lieut. W. C. Cooper, 53rd Sikhs, F.F. ...	30/3/15	23/6/15	—	—
2/Lieut. A. F. Joseph, I.A.R.O.	31/3/15	26/4/15	Wounded	—
Lieut. F. G. C. Campbell, 40th Pathans ...	4/4/15	18/4/15	—	—
Capt. T. Luck, 67th Punjabis	6/4/15	3/2/16	To 67th Punjabis	—
Capt. G. A. Philips, I.A.R.O.	13/4/15	14/6/15	To 129th Baluchis ...	—
Lieut. F. A. Robertson, I.A.R.O.	13/4/15	6/5/15	—	—
Lieut. R. H. Burne, I.A.R.O.	19/4/15	16/4/19	Demobilized	—
2/Lieut. N. de C. Hardwick, I.A.R.O. ...	28/4/15	6/5/15	—	—
2/Lieut. M. Grey-Smith, I.A.R.O.	30/4/15	6/5/15	—	—
2/Lieut. G. Anscombe, I.A.R.O.	13/6/15	17/7/15	—	—
Lieut. H. Steadman	16/7/15	—	—	—
2/Lieut. W. H. H. Young, I.A.R.O. ...	24/9/15	9/1/18	Wound pension	Adjutant, 1916-17.
2/Lieut. B. Moody, I.A.R.O.	24/9/15	27/11/15	—	—
2/Lieut. J. S. Culverwell	13/10/15	25/5/22	Died at Simla	—
Capt. L. M. Heath, 19th Punjabis	7/11/15	8/3/16	Wounded	—
2/Lieut. J. D. Twinberrow	1915	1922	Retired under Royal Warrant	—

NAME, WITH RANK ON JOINING.	DATE OF APPOINTMENT OR JOINING.	BECAME NON-EFFECTIVE.	CAUSE.	REMARKS.
2/Lieut. R. Davis	1915	11/1/17	Died of wounds	—
2/Lieut. E. D. D. Jarrad, I.A.R.O. ...	1915	1917	To recruiting staff ...	—
2/Lieut. G. G. Hills	1915	—	—	—
2/Lieut. A. Morrison	25/12/15	9/8/17	—	—
2/Lieut. R. S. Dudley, I.A.R.O.	25/12/15	18/3/19	Demobilized	—
2/Lieut. J. H. Manley, I.A.R.O.	13/1/16	9/1/17	Killed in action	—
2/Lieut. J. R. Milligan, I.A.R.O.	19/1/16	8/3/16	Killed in action	—
Capt. J. B. Gordon, 52nd Sikhs, F.F. ...	28/1/16	8/3/16	Wounded	—
2/Lieut. L. G. Burgess, I.A.R.O.	28/1/16	15/6/16	—	—
2/Lieut. C. B. Wilson	1916	—	—	—
2/Lieut. C. M. Malden	1916	—	—	—
Lieut. H. L. Cooper, I.A.R.O.	22/3/16	1920	Demobilized	—
Lieut. A. P. Rodgerson, 1/39th Garhwalis	22/3/16	28/7/16	—	—
2/Lieut. E. J. K. Garthwaite, I.A.R.O. ...	27/3/16	22/6/18	Demobilized	—
2/Lieut. C. E. Harris, I.A.R.O.	1916	1921	Invalided	—
2/Lieut. C. W. Chapman, I.A.R.O. ...	4/4/16	7/5/16	—	—
Capt. W. H. Miller, 74th Punjabis	4/4/16	18/4/16	Killed in action	—
2/Lieut. G. E. Hansen	1916	1920	To 2/55th Rifles	Adjutant, 1917-20.
Lieut. F. B. Roseveare	2/9/16	9/11/17	Died of wounds	—
2/Lieut. J. T. B. Bookey, 52nd Sikhs, F.F.	24/12/16	–/9/18	To 6th Gurkha Rifles ...	—
2/Lieut. S. Clapham, I.A.R.O.	29/1/17	26/4/17	—	—
2/Lieut. D. M. Lindsay	30/1/17	—	—	Adjutant, 1921-25.
Major R. D. Beadle, 46th Punjabis ...	15/5/17	2/7/18	To 2/151st Infantry ...	—
2/Lieut. J. J. Jagger, I.A.R.O.	1917	—	—	—
Lieut. D. J. Brown, I.A.R.O.	10/11/17	18/5/20	Demobilized	—
2/Lieut. P. T. Clarke	1917	1918	To 51st Sikhs, F.F. ...	—
Lieut. J. L. Carter, 15th Sikhs	14/11/17	20/3/19	To 54th Sikhs, F.F. ...	—
Lieut. G. H. Simms	30/11/17	1922	Retired under Royal Warrant	—
Lieut. J. M. Stirling	1917	1920	Retired	—
2/Lieut. J. St. C. Arbuthnot	1917	1918	To 2/151st Infantry ...	—
Capt. A. J. Hobley	1917	1922	Retired under Royal Warrant	—
Lieut. H. Buchanan, 2/35th Sikhs	19/6/18	16/4/19	Demobilized	—
Lieut. W. R. C. A. Brook, 54th Sikhs, F.F.	2/7/18	11/1/19	Invalided	—
2/Lieut. J. B. P. Seccombe	1918	—	—	—
Lieut. A. J. Bush, I.A.R.O.	2/7/18	21/7/20	Demobilized	—
Lieut. H. R. Cox, I.A.R.O.	22/7/18	24/7/18	To 56th Rifles, F.F. ...	—
Lieut. J. J. Waite	1918	1919	To 10th Jats	—

Lieut. A. F. Telfer, 2/56th Rifles	3/8/18	16/4/19	Demobilized		—
2/Lieut. H. W. J. Wilkins, I.A. (T.C.) ...	1918	1919	To 14th K.G.O. Sikhs ...		—
Lieut. F. W. Buckler, I.A.R.O.	20/9/18	21/8/20	Demobilized		—
2/Lieut. C. W. Cotton, I.A.R.O.	1918	1919	Demobilized		—
Lieut. R. E. Wilson	1918	1920	To civil employ		—
Lieut. H. N. Lindsay	1918	1922	Retired under Royal Warrant		—
2/Lieut. W. H. Gill, I.A. (T.C.)	1919	1919	Demobilized		—
2/Lieut. T. F. S. Sawyer	1919	1923	Retired under Royal Warrant		—
Lieut. J. A. Robinson	1919	—	—		—
Lieut. G. D. Florence	1919	1922	Retired under Royal Warrant		—
2/Lieut. F. Ashcroft	1920	—	—		—
Lieut. J. W. Banks	1920	1921	Demobilized		—
Lieut.-Col. C. R. Wilkinson	1921	1925	To Unemployed List ...	C.O., 1921-25	
Capt. H. C. Duncan	1921	—	—		—
Lieut. W. H. Fitzmaurice	1921	—	—		—
Capt. D. Bainbridge	1921	—	—		—
Capt. J. G. Wainwright	1921	—	—		—
Capt. N. L. Mitchell-Carruthers	1921	—	—		—
Lieut. W. R. Darnell	1921	1922	Retired under Royal Warrant		—
Lieut. P. C. Evans	1921	1922	Retired under Royal Warrant		—
2/Lieut. H. G. L. Brain	1921	—	—		—
Brevet Lieut.-Col. A. E. Mahon	1922	1923	To 1/12th F.F. Regt. ...		—
Lieut. S. F. J. Hodgman	1922	1923	Retired under Royal Warrant		—
Capt. G. A. P. Coldstream	1922	—	—	Adjutant, 1925	
Capt. D. H. J. Williams	1923	—	—		—
Major W. E. H. Condon	1923	—	—		—
Capt. G. T. Pearson	1924	—	—		—

GENERAL OFFICERS COMMANDING THE 8TH JULLUNDUR BRIGADE, 1914–1920.

Major-General P. M. Carnegy, C.B. ... ... August–December, 1914.
Brigadier-General E. P. Strickland, C.M.G., D.S.O. ... ... January–November, 1915
Brigadier-General S. M. Edwardes, C.B., C.M.G., D.S.O. ... ... 1916–1920.

APPENDIX IV

Copy of Farewell Order issued to the 59th Rifles F.F. by Major-General Keary, on relinquishing Command of the 3rd Lahore Division, in September, 1917.

FAREWELL ORDER TO THE 59th RIFLES.

On relinquishing command of the 3rd Division, I desire to add a special word of praise and thanks to the 59th Rifles. Your Regiment has vied with the 1st Manchesters and the 47th Sikhs in making the 8th Brigade what it is, and where all have done so well it is difficult to select one.

In all the great battles the Division has fought during these last three years, your Regiment has taken its part most worthily, affording the most loyal co-operation and support to neighbouring units at all times.

One of the most marked characteristics of the Regiment has been its unselfish sacrifice in assisting others in difficulties.

I have always had the fullest confidence in you under all circumstances, and not once have you failed to distinguish yourselves when called upon.

Your conduct and discipline have been remarkable, and it would be difficult to find in the Indian Army a Regiment that sets a higher standard for fighting or marching ability.

I tender you my best thanks for your gallant behaviour and your loyal support at all times.

I wish you the best of good fortune in the future, and look forward to serving with you again one day.

(Signed) H. D. Keary,
Major-General.

APPENDIX V

Decorations received during the Great War, 1914-1918

REGTL. NO.	RANK.	NAME.	DECORATIONS.
—	Lieutenant	W. A. McC. Bruce	V.C.
—	Lieut.-Colonel	T. L. Leeds	C.M.G.
—	Lieut.-Colonel	T. L. Leeds	D.S.O.
—	Captain	J. R. Wynter, 52nd Sikhs, F.F.	D.S.O.
—	Major	B. E. Anderson	D.S.O.
—	Captain	R. D. Inskip	D.S.O.
—	Captain	K. D. Murray	D.S.O.
—	Major	A. H. Burn	O.B.E.
—	Lieutenant	J. A. M. Scobie	M.C.
—	Captain	R. D. Inskip	M.C.
—	Captain	H. F. D. Stirling	M.C.
—	Captain	J. G. B. Gordon, 52nd Sikhs, F.F.	M.C.
—	Captain	L. M. Heath, 19th Punjabis	M.C.
—	Captain	R. H. Burne, I.A.R.O.	M.C.
—	Subadar-Major	Perbhat Chand	M.C.
—	Lieutenant	C. H. N. Baker, I.M.S.	M.C.
—	Lieutenant	R. N. Kapadia, I.M.S.	M.C.
—	Lieutenant	J. D. Twinberrow	M.C.
—	Lieutenant	W. H. Young	M.C.
—	Captain	A. F. Telfer, 2/56th Rifles, F.F.	M.C.
—	Lieutenant	J. L. Carter, 15th Sikhs	M.C.
—	Subadar-Major	Nasir Khan	Order of British India, 1st Class.
—	Subadar	Makhmad Jan	Order of British India, 1st Class.
—	Subadar	Gauri Charan	Order of British India, 2nd Class.
—	Subadar-Major	Nasir Khan	I.O.M., 2nd Class.
3907	Sepoy	Biaz Gul	I.O.M., 2nd Class.
27*	Sepoy	Zarif Khan	I.O.M., 1st Class.
3063	Havildar	Abdul Wahab	I.O.M., 2nd Class.
3191	Havildar	Dost Mohammed	I.O.M., 2nd Class.
3638	Havildar	Mohd. Jan	I.O.M., 2nd Class.
3663	Havildar	Muzaffar Khan	I.O.M., 2nd Class.
3705	Lance-Naik	Buta Singh	I.O.M., 2nd Class.
—	Subadar	Khan Gul	I.O.M., 2nd Class.
—	Subadar	Gauri Charan	I.O.M., 2nd Class.
4040	Havildar	Qalandar Khan	I.O.M., 2nd Class.
2551	Havildar	Sunit Chand, 52nd Sikhs, F.F.	I.O.M., 2nd Class.
1194	Lance-Naik	Lal Khan, 19th Punjabis	I.O.M., 2nd Class.
2665	Naik	Ghulam Hassan, 52nd Sikhs, F.F.	I.O.M., 2nd Class.
—	Subadar	Tola Singh	I.O.M., 2nd Class.
4551	Lance-Naik	Ali Faqir	I.O.M., 2nd Class.

* Sepoy Zarif Khan gained the 2nd Class I.O.M. in 1914, and in May, 1915, was awarded the 1st Class I.O.M. for gallantry in action at Neuve Chapelle.

REGTL. NO.	RANK.	NAME.	DECORATIONS.
2549	Sepoy ...	... Kahn Singh, A.N.M.P. ...	... I.O.M., 2nd Class.
—	Subadar ...	... Sahib Haq	... I.O.M., 2nd Class.
—	Subadar-Major	... Mohd. Khan	... I.D.S.M.
4264	Havildar ...	... Niaz Gul	... I.D.S.M.
4731	Sepoy	... Lal Khan	... I.D.S.M
—	Jemadar ...	... Zaman Ali	... I.D.S.M.
3524	Naik ...	... Amir Ali	... I.D.S.M.
2520	Lance-Naik	... Chur Khan, 52nd Sikhs, F.F.	... I.D.S.M.
3581	Naik ...	... Ghamai Khan	... I.D.S.M.
4805	Sepoy ...	... Akbar Khan	... I.D.S.M.
—	Subadar ...	... Bishen Singh	... I.D.S.M.
2511	Havildar ...	... Mir Afzal, 25th Punjabis ...	... I.D.S.M.
4770	Lance-Naik	... Phuman Singh	... I.D.S.M.
1955	Havildar ...	... Sohnu Ram, 52nd Sikhs, F.F.	... I.D.S.M.
3086	Lance-Naik	... Ranjha, 52nd Sikhs, F.F. ...	... I.D.S.M.
4111	Havildar ...	... Nathu	... I.D.S.M.
2099	C. Havildar	... Kapur Chand, 52nd Sikhs, F.F.	... I.D.S.M.
4016	Havildar ...	... Nur Ali	... I.D.S.M.
1723	Lance-Naik	... Madar Khan, 56th Rifles, F.F.	... I.D.S.M.
—	Subadar ...	... Hakam Khan, 25th Punjabis	... I.D.S.M.
2361	Havildar ...	... Kajir Khan, 25th Punjabis	... I.D.S.M.
2266	Havildar ...	... Indar Singh, 52nd Sikhs, F.F.	... I.D.S.M.
3590	Havildar ...	... Mit Singh	... I.D.S.M.
3601	Lance-Naik	... Saida Khan	... I.D.S.M.
1164	S.A.S. ...	... Chuhar Singh, I.M.D. ...	... I.D.S.M.
2949	Naik ...	... Sunder Singh, 52nd Sikhs ...	... I.D.S.M.
3452	Havildar ...	... Devi Dyal	... I.D.S.M.
4342	Havildar ...	... Niaz Gul	... I.D.S.M.
2696	Naik ...	... Dhani Ram, 52nd Sikhs, F.F.	... I.D.S.M.
4183	Havildar ...	... Hazrat Khan	... I.D.S.M.
4116	Havildar ...	... Abuzar Khan	... I.D.S.M.
—	Subadar ...	... Ali Akbar	... I.D.S.M.
—	Jemadar ...	... Malang Khan	... I.D.S.M.
4813	Havildar ...	... Hukam Dad	... I.D.S.M.
150	Naik ...	... Alawal Khan	... I.D.S.M.
357	Sepoy ...	... Hasham Khan	... I.D.S.M.
4902	Naik ...	... Rasul Khan	... I.D.S.M.
2210	Naik ...	... Mir Alam Khan, 56th Rifles, F.F. ...	I.M.S.M.
4158	Naik ...	... Mohd. Ramzan	... I.M.S.M.
4555	Naik ...	... Panni Lal	... I.M.S.M.
4537	C. Havildar	... Parru	... I.M.S.M.
3228	C. Havildar	... Punjab Singh, attd. 112th C.F.A. ...	I.M.S.M.
1030	K. Daffr. ...	... Zainul Abdin, S. & T. Corps	... I.M.S.M.
1199	Sepoy ...	... Wazir Chand	... I.M.S.M.
4902	Naik ...	... Rasul Khan	... I.M.S.M.
315	Havildar-Major	... Makhmad Yusaf	... I.M.S.M.
3915	Naik ...	... Nizam Din	... I.M.S.M.
47	Havildar ...	... Akbar Khan	... I.M.S.M.
4724	Havildar ...	... Dilbar Shah	... I.M.S.M.
3929	Havildar ...	... Wali Jan	... I.M.S.M.
150	Havildar ...	... Alawal Khan	... I.M.S.M.
2825	Naik ...	... Sarwan Singh, A.N.M.P. ...	... I.M.S.M.
2392	Havildar ...	... Gokal, 52nd Sikhs, F.F. ...	... I.M.S.M.
4259	Havildar ...	... Anup Singh	... I.M.S.M.
98	Lance-Naik	... Santa Singh	... I.M.S.M.

REGTL. NO.	RANK.	NAME.	DECORATIONS.
3991	Sepoy	Ran Singh	I.M.S.M.
4770	Havildar	Phuman Singh	I.M.S.M.
1209	Sepoy	Lal Singh	I.M.S.M.
4354	Naik	Alam Gul	I.M.S.M.
46	Naik	Sohbat Khan	I.M.S.M.
4194	Havildar	Ram Ditta	I.M.S.M.
94	Havildar	Diwan Ali	I.M.S.M.
4513	Havildar	Fazal Ilahi	I.M.S.M.
4530	Havildar	Mir Mast	I.M.S.M.
310	Naik	Allah Nur	I.M.S.M.
3849	Havildar	Ram Singh	I.M.S.M.
2981	Naik	Ghazan Khan, 52nd Sikhs, F.F. ...	I.M.S.M.
167	Sepoy	Lal Khan	I.M.S.M.
3651	Havildar	Sowan Singh	I.M.S.M.
1	Sepoy	Fakir Singh	I.M.S.M.
14	Sepoy	Kirpa Singh	I.M.S.M.
3625	Havildar	Amir Khan	I.M.S.M.
479	Naik	Mashal Khan	I.M.S.M.
4204	Naik	Makhmad Gul	I.M.S.M.
4438	Lance-Naik ...	Sundar Singh	I.M.S.M.
4925	Lance-Naik ...	Ghasita	I.M.S.M.
4719	Sepoy	Sher Singh	I.M.S.M.
16	Sepoy	Sohan Singh	I.M.S.M.
5	Sepoy	Subha	I.M.S.M.
144	Sepoy	Lalmai	I.M.S.M.
694	Sepoy	Mir Alam	I.M.S.M.
4024	Naik	Budh Singh	I.M.S.M.
312	Naik	Buta Singh	I.M.S.M.
193	Sepoy	Teja Singh	I.M.S.M.
4363	Havildar	Fazal Dad	I.M.S.M.
151	Lance-Naik ...	Tahaur Khan	I.M.S.M.
336	Lance-Naik ...	Firoz Khan	I.M.S.M.
4155	Naik	Ladha Khan	I.M.S.M.
3955	Havildar	Gurditta	I.M.S.M.
4590	Naik	Sohan Singh	I.M.S.M.
4658	Lance-Naik ...	Sohan Singh	I.M.S.M.
126	Sepoy	Udham Singh	I.M.S.M.
140	Sepoy	Mahi Singh	I.M.S.M.
1599	Sepoy	Gujjar Singh	I.M.S.M.
4666	Lance-Naik ...	Narain Singh	I.M.S.M.
4638	Havildar	Aman Ali	I.M.S.M.
4738	Lance-Naik ...	Mohd. Hussain	I.M.S.M.
3745	Havildar	Sajawal Khan	I.M.S.M.
4154	Lance-Naik ...	Niaz Ali	I.M.S.M.
198	Sepoy	Sultan Khan	I.M.S.M.
4199	Havildar	Akbar Khan	I.M.S.M.
3990	Havildar	Niaz Khan	I.M.S.M.
3229	Havildar	Faujdar Khan	I.M.S.M.
1297	Sepoy	Shankar Singh	I.M.S.M.
3312	Sepoy	Nur Mohd. (ward orderly attd. 112th C.F.A.)	I.M.S.M.
3756	Havildar-Major ...	Karm Singh	I.M.S.M.
3454	Havildar	Shah Yusaf	I.M.S.M.

FOREIGN DECORATIONS.

RUSSIAN DECORATIONS.

REGTL. NO.	RANK.	NAME.	DECORATIONS.
—	Subadar-Major ...	Nasir Khan	Medal of St. George, 1st Class.
3063	Havildar	Abdul Wahab	Medal of St. George, 2nd Class.
3890	Sepoy	Saddar Din (ward orderly)	Medal of St. George, 4th Class.
—	Bt. Lieut.-Colonel	T. L. Leeds	Order of St. Anne, 3rd Class (with Sword).
2665	Naik	Ghulam Hassan, 52nd Sikhs, F.F.	Medal of St. George, 1st Class.
2270	Lance-Naik ...	Gopal Singh, 52nd Sikhs	Medal of St. George, 3rd Class.
		ITALIAN DECORATION.	
4927	Lance-Naik ...	Hassan Ali	Bronze Medal for "Military Valour."
		FRENCH DECORATION.	
631	Lance-Naik ...	Shahbaz Khan	Medaille Militaire.
		SERBIAN DECORATION.	
3601	Havildar	Saida Khan	—
		MISCELLANEOUS.	
—	Corporal	Paul Ries (French interpreter) ...	D.C.M.
		BREVETS.	
—	Major	T. L. Leeds	Bt. Lieut.-Colonel.
—	Captain	B. E. Anderson	Brevet Major.
—	Captain	R. D. Inskip	Brevet Major.

GOVERNMENT OF INDIA LAND AWARDS AND "JANGI INAMS."

The Regiment received a total of two hundred and forty-one of these awards under the Government of India War Reward Scheme of 1920. These included two Special Jagirs granted to :—

Subadar-Major (Hon. Captain) Nasir Khan Sardar Bahadur, I.O.M., and Subadar Gauri Charan, Bahadur, I.O.M.

MENTIONS IN DESPATCHES, 1914–1918.

FRANCE.

Captain B. E. Anderson.
Captain H. N. Lee.
4264 Havildar Niaz Gul.
Lieut.-Colonel P. C. Eliott-Lockhart.
Major T. L. Leeds.
Lieut. Colonel C. C. Fenner.
Captain T. Reed (67th Punjabis, attd.).
Lieutenant J. A. M. Scobie.
Subadar Perbhat Chand.
Captain P. S. Hore (52nd Sikhs, F.F., attd.).
Captain R. D. Inskip.
Major K. D. B. Murray (twice).

MESOPOTAMIA.

Captain (A./Lieut.-Colonel) H. F. D. Stirling, M.C. (twice).
Lieut. W. H. H. Young, I.A.R.O.
Subadar Bahadur Shah.
4911 Havildar Nathu.
4342 Havildar Niaz Gul.
4278 Naik Mir Hasham.
4462 Naik Kasir Khan.
4402 Sepoy Hamam Singh.
Captain C. H. N. Baker, I.M.S.
Lieutenant R. S. Dudley, I.A.R.O.
3452 Havildar Deri Dyal.
Captain A. H. Burn.
Lieut.-Colonel T. L. Leeds (twice).
4199 Naik Akbar Khan.
Captain R. H. Burne, I.A.R.O.
2266 Havildar Inder Singh (52nd Sikhs, F.F., attd.).
2989 Naik Sundar Singh (52nd Sikhs, F.F., attd.).
Brevet Major B. E. Anderson.
Captain R. D. Inskip.
Captain L. M. Heath (19th Punjabis, attd.).
Captain J. G. B. Gordon (52nd Sikhs, F.F., attd.).
Jemadar Sahib-i-Haq.
Subadar Thola Singh.
4551 Lance-Naik Ali Faqir.
4104 Naik Gandarab Singh.
3590 Havildar Mit Singh.
3601 Lance-Naik Saida Khan.
1164 S.A.S. Chur Singh.
Subadar Abdullah Shah (19th Punjabis, attd.).
1194 Lance-Naik Lal Khan (19th Punjabis attd.).
1635 Sepoy Said Abbas (19th Punjabis, attd.).
Subadar Hakam Khan (25th Punjabis, attd.).
2361 Havildar Kajar Khan (25th Punjabis attd.).
2217 Havildar Bahadur Shah (52nd Sikhs, F.F., attd.).
2265 Naik Ghulam Hussein (52nd Sikhs, F.F., attd.).

PALESTINE.

Major B. E. Anderson.
Lieut.-Colonel T. L. Leeds.
4902 Naik Rasul Khan.

SERVICES IN INDIA.

Jemadar Karm Din.

The following distinctions were gained in operations after the Great War :—

MESOPOTAMIA REBELLION.

C.I.E., Mentioned in Despatches—Major A. H. Burn.

SOMALILAND, 1919–20.

I.D.S.M., I.M.S.M., Mentioned in Despatches—Subadar Burhar Ali Khan.

RAZMAK FIELD FORCE.

Mentioned in Despatches—Major A. H. Burn, C.I.E., O.B.E., Major B. E. Anderson, D.S.O.,

APPENDIX VI

Honorary Distinctions

In recognition of the distinguished services and gallantry of the Indian Army during the Great War, His Majesty the King Emperor has been graciously pleased to confer the title "Royal" on the undermentioned units :—

20/29th Deccan Horse.
3rd Sappers and Miners.
6th Jat Light Infantry.
34th Sikh Pioneers.
39th Garhwal Rifles.
59th Scinde Rifles (Frontier Force).
117th Mahrattas.
5th Gurkha Rifles (Frontier Force).

His Majesty has also been pleased to nominate Field-Marshal His Royal Highness Arthur William Patrick Albert, Duke of Connaught and Strathearn, K.G., K.T., K.P., G.C.B., G.C.S.I., G.C.M.G., G.C.I.E., G.C.V.O., G.B.E., A.D.C., as Colonel-in-Chief, 47th Sikhs, in recognition of its distinguished services and gallantry during the War.

—*Indian Army Order* 821, *dated* 26/7/21.

www.ingramcontent.com/pod-product-compliance
Ingram Content Group UK Ltd.
Pitfield, Milton Keynes, MK11 3LW, UK
UKHW050146280726
14058UKWH00007B/862